PRECISION FARMING

Soil Fertility and Productivity Aspects

PRECISION FARMING

Soil Fertility and Productivity Aspects

K. R. Krishna, PhD

Apple Academic Press

TORONTO NEW JERSEY

© 2013 by
Apple Academic Press Inc.
3333 Mistwell Crescent
Oakville, ON L6L 0A2
Canada

Apple Academic Press Inc.
9 Spinnaker Way, Waretown, NJ 08758
USA

First issued in paperback 2021

Exclusive worldwide distribution by CRC Press, a Taylor & Francis Group

ISBN 13: 978-1-77463-270-3 (pbk)
ISBN 13: 978-1-926895-44-4 (hbk)

Library of Congress Control Number: 2012951943

Library and Archives Canada Cataloguing in Publication

Krishna, K. R. (Kowligi R.)
Precision farming: soil fertility and productivity aspects/K.R. Krishna.

Includes bibliographical references and index.
ISBN 978-1-926895-44-4
1. Precision farming. 2. Soil fertility. 3. Soil productivity.
I. Title.

S494.5.P73K75 2013 631 C2012-906399-1

About the Author

K. R. Krishna, PhD

K. R. Krishna received his PhD in agriculture from the University of Agricultural Sciences in Bangalore. He has been a cereals scientist in India and a visiting professor and research scholar at the Soil and Water Science Department at the University of Florida, Gainesville, USA. Dr. Krishna is a member of several professional organizations, including the International Society for Precision Agriculture, the American Society of Agronomony, the Soil Science Society of America, the Ecological Society of America, and the Indian Society of Agronomy.

Contents

Contents

List of Abbreviations

AE_N	Agronomic efficiency of fertilizer-N
BMP	Best management practice
CEC	Cation exchange capacity
DGPS	Differential global positioning system
EONR	Economically optimum rates of N
EC	Electrical conductivity
EMI	Electromagnetic induction
FYM	Farmyard manure
GIS	Geographic information system
GPS	Global positioning systems
INM	Integrated nutrient management
LED	Light emitting diode
NIR	Near-infrared
NDVI	Normalized difference vegetative index
NMR	Nuclear magnetic resonance
PF	Precision farming
RVI	Reflectance vegetation index
SAR	State agency recommendation
SSNM	Site-specific nutrient management
SOC	Soil organic carbon
SOM	Soil organic matter
SPAD	Soil plant analysis development
STCR	Soil test crop response
VRT	Variable rate technology
WFM	Whole farm management

Preface

Precision is an important concept that has induced, guided, and improved many aspects of human endeavor for ages. Historically, precision has been an important factor in agricultural evolution *per se*. Precision in planting dates and matching crops with seasons to derive maximum advantage from soils and precipitation patterns are perhaps the oldest aspects of agriculture. Farmers have been striving to achieve greater accuracy on these aspects of agricultural cropping since the Neolithic period. They continue to do so even today. There is no doubt that through the ages, precision as a concept has been imbibed into almost every technique and practice during crop production. Precision has been a key factor in selection of healthy seeds, seeding technique, fertile locations, types of manures, and moist zones. Precision has sometimes caused quantum changes in cropping pattern and productivity. Precision in matching soil fertility and its variations with crop species or its genotype with yield goals is a key aspect of agriculture in any part of the world. Precision is needed while selecting a crop genotype. The genotype should match the agro-environment, soils, season, grain yield goals, and profitability. Agricultural cropping trends and productivity, in particular, have depended on the extent of precision bestowed on farming procedures. For example, precise crop genotype, accurate supply of nutrients, and irrigation in time and space were major factors in improving crop productivity during first half of the 20th century. Today, precision techniques offer farmers the greatest opportunity to regulate soil nutrient dynamics, protect agro-environment and yet enhance crop productivity.

During recent years, a perceptibly greater degree of precision has been incorporated into almost all farming procedures. Soil fertility and manure supply trends, in particular, have received greater attention with regard to the extent of precision possible. The advent of computer models, simulations, and decision support systems have allowed us to direct exact quantities of seeds, fertilizers, water, and pesticides through the use of variable-rate technology. Actually, equipments such as computers, handheld sensors, and satellite-guided systems have remarkably enhanced precision during farming. Precision technique creates uniform soil fertility across a field. Grain/fruit and forage productivity too become uniform commensurately. Precision techniques often envisage use of slightly or markedly lower quantities of fertilizers and irrigation to achieve same levels of crop productivity. The reduction in fertilizer usage delays or totally avoids deterioration of soils, ground water, aquifers, and general agro-environment. Precision techniques also provide higher profits to farmers. Overall, reduction in use of natural resources, improved grain/forage yield, and extra profitability compared to farmer's traditional procedures hold the key to its rapid acceptance in most agricultural regions of the world.

During past decade, rapid improvements have occurred in precision techniques. Improvization of GPS-guided farm machinery, sensors, data capture, soil fertility mapping procedures, and GPS-guided variable-rate techniques have been marked.

The spread of precision techniques into different agricultural belts and evaluation of its advantages have received the greatest attention. Precision technique is most recent among the agronomic procedures exposed to farmers/researchers. Field evaluations across different continents suggest it could be a very popular and profitable technology in the near future.

This book, titled *Precision Farming*, is introductory. It begins with a discussion on historical aspects, provides brief descriptions on techniques, and enlists advantages as well as constraints that influence the adoption and spread of precision farming in different continents. Chapter 2 provides details on intricate instrumentation, their functioning, and advantages that accrue during precision farming. Chapter 3 forms the centerpiece of this book. It deals with the influence of precision farming approaches on soil fertility, nutrient dynamics, and productivity of various crops. The spread of precision farming methods into different geographic regions and profitability are discussed in detail in Chapter 4. A brief discussion about the future course of precision farming approaches appears in the last chapter.

This book on precision techniques is concise and provides valuable information on instrumentation and methodology. It encompasses lucid discussions about the impact of precision techniques on soil fertility, nutrient dynamics, and crop productivity. It is most useful to students, researchers, and professors involved in various aspects of agriculture.

— K. R. Krishna, PhD

Acknowledgments

Dr. Eric Lund and others of Veris Technologies, Salina, Kansas, USA, provided pictures of sensors that estimate soil pH, electrical conductivity, and nutrients. Pictures on precision techniques such as management strips, strip tillage, variable-rate supply of fertilizer and seeds, GPS-guided seeder, fertilizer mixing trucks, and fertilizer application systems were obtained from Mr. David Nelson, Nelson Farms Inc, Fort Dodge, Iowa, USA. Pictures of hand-held portable instruments that estimate photosynthetic activity were obtained from Mr. Micheal Larman, CID-BIO Science, Camas, Washington, USA. Pictures of hand-held sensors that measure Leaf-N and chlorophyll content, help in gauging N status of a crop, and in forecasting yield were derived from Konica-Minolta Sensing Inc, New Jersey, USA.

I wish to thank my wife Dr. Uma Krishna and son Mr. Sharath Kowligi.

1 Introduction

CONTENTS

1.1 HISTORICAL ASPECTS

Precision means exactness or accuracy in performance of a particular task. In the present context, it refers to accuracy of various agricultural practices and farming per square that are carried out by farmers. Precision is actually a concept that got imbibed into agricultural endeavor of human beings, since pre-historic times. Precision also induced evolution of agricultural techniques and farming. Earliest of the steps towards precision could be seen in preferential seeding of a particular crop species in the vicinity of prehistoric human dwelling sites of early to late Neolithic period. Neolithic farmers gained by seeding and growing a precise crop species in those backyards of their dwellings. It overcame difficulty in tedious collection of grains from swamps or plains that had admixtures of all kinds of plant species. In this case, precision in terms of crop species and domestication allowed farmer greater quantity of harvests. It is interesting to note that since these early stages of agricultural history, precision as a concept has been quietly imbibed and utilized too, mainly to make farming easier and enhance productivity of land. Farmers devised procedures and manufactured implements that enhanced accuracy. Precise soil management using plough meant better nutrient and water management. Farmers introduced ploughing and line sowing during ancient period. Ploughing and line sowing is indeed a conspicuous effort to add greater degree of precision into farming, in terms of planting geometry and density, efficient interception of light, as well as moisture and nutrient scavenging. It is a major event in the agricultural history that added precision to farms worldwide. Line sowing improved crop production compared to a field randomly broadcasted with seeds. Tillage and line sowing added precision to several other procedures like timing of interculture operations, top dressing, irrigation, pesticide application, and harvesting. During modern era (1820th century), precision got imbibed into agriculture through various improvements that farmers effected on to their implements, seeding procedures, irrigation devises, harvesting, and grain processing. During this period of history, farmers literally gained in efficiency and productivity by adding precision to farming procedures. Most glaring of the procedures introduced by farmers that added precision are precise planting dates to match with precipitation pattern and season. Even today, we strive hard to add precision into planting dates, seeding depth and plant population because it improves nutrient scavenging, moisture absorption and grain harvests significantly. Precise crop species and precise field to match fertility requirements of crops are other measures that improved productivity. Irrigation channels helped farmers in the supply of precise quantities of water at various stages of the crop development. During recent decades, there has been a steady improvement in precision aspects of implements, gadgets and procedures adopted in the field. Invention of fertilizer formulations improved accuracy further. Soil fertility could be mended accurately and sustained despite repeated cropping of the same field. Soil nutrients could be accurately replenished using various soil chemical analysis procedures and soil test crop response (STCR) studies. Automatic irrigation based on periodic soil moisture measurements and crops' need improved accuracy of crop production. Together, aspects like precise crop species/genotype, nutrient replenishment and irrigation were instrumental in enhancing crop yield. We should note that precise selection and genetic improvement of crop species has added to grain harvests significantly.

Historically, selection of genotypes that flowered and matured uniformly, produced non-dehiscent panicles/seeds of uniform traits have added to accuracy. Production of genotypes with uniform height and panicle development (semi-dwarfs) that aided efficient mechanical harvest is a glaring example that depicts gain in precision through crop breeding. Since mid 1900s, precision as a concept was imbibed as a matter of routine into almost all aspects of farming. Throughout past decades, almost every modification in traction machinery, types of coulters, their shape, size, seed drills, fertilizer drills, hoes, weeders, harvesters, right up to development of elaborate combine harvesters have all aimed at enhancing accuracy of specific tasks. Gadgets driven mechanically or through electrical power added further to precision of agricultural techniques. Electronic controls and timing too added precision to various farm operations.

Historically, most recent of the farming measures that seems to add precision into farming procedures, yet again and in significant amount, is the use of satellite-guided seeding, fertilizer application, irrigation, pesticide spray, harvesting, and yield monitoring. Satellite-guided procedures and computer-aided decision support systems are forecasted to revolutionize the way agricultural farms are managed. Such procedures are collectively termed Precision Farming (PF) because they add accuracy to soil fertility and moisture management, rather enormously. In summary, precision is a concept that has been imbibed into farming. It has helped farmers to carry out various tasks efficiently with due gains in input efficiency and grain/forage harvests. Right now, we have no idea regarding limit to precision or accuracy in farming procedures and the extent of benefits it may fetch. We ought to appreciate that evolution of agricultural operations and gain in precision has affected nutrient dynamics and productivity of the land, either directly or inadvertently. This aspect should be needs to understand in greater detail.

The PF as we know today currently involves, remote sensing, Global Positioning Systems (GPS) guided instrumentation, fertilizer supply based on Geographic Information System (GIS), computer models, and variable rate nutrient applicators. Historically, this aspect of farming is only one and a half decade old. Its development, refinement in technology, introduction and rapid spread into different cropping zones has occurred since mid 1990s. It is relatively a new scientific aspect with a short stretch of history. Some of the earliest references to within field variability pertaining to soil moisture, nutrients, and pH were made as early as 1986. Actually, precision agriculture as a concept that overcomes within field variability in soil fertility and one that provides better synchronization between nutrient need and supply was envisaged in 1986 (Fairchild, 1994). It received greater attention as a method that has impact on resource use efficiency, crop production, profitability, and environment during 1990s (Earl et al., 1996; Gerhards et al., 1996; Khakural et al., 1996).

Farmers situated in most regions of the world, including highly intensive crop production zones like Corn Belt of USA or Wheat expanses in Europe or rice cultivation areas in Southeast Asia or subsistence farming zones in semi-arid regions were ordinarily accustomed to simple and traditional techniques to estimate soil fertility. Then, they prescribed fertilizers/organic manure and recorded harvest derived from entire field. Identification of within field variations regarding soil physico-chemical properties, fertility trends and obtaining productivity maps was not practiced routinely. According to Berry et al. (2010), soil or yield maps played insignificant roles in crop production until mid 1990s. Soil and topographical maps were more generalized and directed towards demarcation of soil fertility regimes. Agricultural crop production was based mostly on whole field estimates of soil fertility. In fact, only broad averages were considered while taking nutrient management decisions. Soil sampling was superficial and done to know physico-chemical conditions and nutrient status at a broad field level. Grain elevators and combine harvesters recorded only final yield per field/farm allowing us to compute only average grain yield per hectare. Such data were used as efficiently and authentically to decide about cropping systems, soil fertility measures, planting densities, irrigation, and disease/pest control measures and harvest schedules. It is interesting to note that during past 15 years, geospatial technology as applied to agricultural crop production has improved enormously. This technology has expanded rapidly from a mere practice to operational reality in several million hectare all over the world (Berry et al., 2010). It is a fact that, at present, even least sophisticated tractors or harvesters bought come with ability to record and map grain yield. Most tractors with advanced instrumentation sense soil fertility variations. They are also equipped with variable rate technology (VRT). Lowenberg-DeBoer (2003a) has made an interesting comparison regarding adoption and spread of PF techniques. He states that initially, pattern of spread of precision techniques was slow and uneven. This seems more like the situation found with motorized mechanization of farm operation in the first half of 20th century or adoption of "No-tillage systems" during second half of 21st century. Hence, spread of precision techniques was not similar to rapid acceptance of say Hybrid Maize in 1930s or hybrids of other cereal species like Sorghum or Triticum. It has been argued that precision techniques are evolving and becoming more efficient with time rather slowly. It is not a finished product unlike a "Hybrid Corn".

Adoption of PF has been relatively more widespread and rapid in developed nations. Farmers have moved from conventional mechanized farming to high technology PF that is guided by GPS, computer based decision support and electronically controlled variable rate applicators and other farm machinery. It is interesting to note that during past decade, sales, and use of equipments necessary for PF such as monitors and variable applicators have increased by 70% in USA and Canada (Tran and Nguyen, 2008). Large farms common to North American agricultural zones seems to induce adoption of PF. Whereas, in the Fareast, small farms seem to make its adoption slower. However, there are clear instances wherein farm size or geographic location does not seem to retard acceptance of PF. It is interesting to note that during initial years in 1990s, the PF techniques were adopted more by farmers situated in proximity to farm and fertilizer dealerships and stores. Farmers with fertile soils and those intending to obtain greater profits from nitrogen fertilizer adopted PF in greater number than, ones who were financially poor and possessed fields with low fertility soils (Kessler and Lowenberg-DeBoer, 1998). In the Mid-west region of USA, a few other surveys directed at assessing farmers adopting PF techniques indicated that in early 1990s, farmers who were young, less than 50 years old and educated up to collegiate level preferred newer techniques like VRT. Farmers previously exposed to computers and those already using electronic gadgets adopted PF despite its elaborate requirement of sampling, GIS, GPS, and VRT instrumentation (Khanna et al., 1999).

Dobermann et al. (2004) have pointed out that initial spurt in adoption of PF in North America and Europe was primarily driven by need to maximize profits, reduce fertilizer-N consumption and NO_3 leaching into ground water. However, adoption of PF was only patchy in North America, Europe, and Australia. Regarding initial use of yield monitors, Lowenberg-DeBoer (2003a) has stated that such instruments were first used in 1992 in USA. Since then its use has grown rapidly, especially in the Great Plains area. In 2003, 30% of tractors used in maize farms were fitted with yield monitors, 25% of soybean fields and 10% of wheat cultivating zones were being assessed using yield monitors. Currently, PF is a popular technology in the Corn Belt and entire Great Plains, where wheat and cotton cropping dominates. Further, it has been pointed out that, in 2004 A.D., about 30,000 yield monitors were in use in USA, 800 in Australia, 800 in Argentina, 1300–1500 in European cereal belt and a few on experimental basis in Asian cropping zones. This is indicative of popularity and usefulness of PF as perceived by farmers. Presently, European cropping zones that support cereal and forage production too use precision techniques more frequently. Dobermann et al. (2004) state that commercialization of soil and crop sensors began in 2000 A.D. It mainly involved estimation of pH, electrical conductivity, and soil nutrients. Whereas, crop sensors develop during this period, usually estimated leaf color, chlorophyll content, and moisture.

Historically, the VRT was among the earliest of site specific technologies adopted on farms. It was used first in 1992 in the "Corn Belt of Untied States of America". It was used to spread fertilizer evenly into farms/fields. Initially, VRT depended on accrued data or maps obtained from grid sampling. The VRT spread rapidly in the mid-west region of USA. In 1996, over 29% of cereal farms adopted VRT and the area improved to 50% in 2002 (Lowenberg-DeBoer, 2003a). Adoption of VRT was generally profitable to farmers in the Great Plains. High value crops responded better than low productivity cereal zones, whenever VRT was adopted to supply fertilizer. Currently, "on-the-go" soil analysis, decision support systems and rapid incorporation of fertilizers is practiced. However, overall goal of VRT is still the same even after 25 years. It aims at finding spatial variations and supplying nutrients/water appropriately at target spots (Sudduth, 2007).

The PF is an apt suggestion in countries that support large sized farms and expansive agriculture, for example in Argentina, Australia, United States of America, some regions in Western Europe, and Russia. PF could be a preferred technique in most of the large sized commercial farms situated anywhere on the globe. It is interesting to note that in USA, farms less than a square mile would become nonviable. Basically, these are large farms and a "field" would mean a vast area with wide fluctuations in soil fertility, especially nutrient and water distribution in the surface and subsurface of soil profile. Blanket application of fertilizer-based nutrient supply creates a mismatch. It results in unequal nutrient removals from patches of soil. Crops tend to be uneven with regard to productivity. Soil fertility continues to stay uneven in the entire farm. Therefore, precision approaches like, closely spaced grid sampling of soils and their analysis is a necessity. Demarcation of large field into "management zone" is a good alternative. Maps depicting soil fertility variations can be used to apply fertilizers accurately resulting in better economics, in addition to providing uniformity to soil fertility. According to Rickman et al. (2003), there are over 2 million farms in USA that are large and need to be managed using precision techniques. The situation is similar with large farms in other regions of the world. The PF allows us to impart uniformity in soil fertility and economize on use of fertilizers and water, especially in agricultural regions with large sized farms. During 2000–2010, extension programs in Ohio State of USA aimed at transmitting knowledge about PF from "early adopters" to those considering its adoption. The program aims at spreading information about usefulness of precision techniques to environmental quality, nutrient dynamics in the farm and grain yields. They mainly at aimed sharing details on yields monitors, variable rate applicators, and information gathering techniques (Batte and Arnholt, 2003).

Economic surveys conducted in the farming zones of South Central Plains, especially in Arkansas, indicated that about 35% rice farmers, 2% each of soybean, cotton, maize, and wheat cultivators had adopted PF in one form or other. Forecasts by Griffin et al. (2000) suggested that fraction of farmers shifting to precision techniques would increase rapidly by 10–20% in case of crops like rice, soybean, and wheat, say in 38 years time. Currently, farmers using VRT has crossed 20% in Arkansas, USA. Adoption of PF was induced through special incentives in several states of southern USA. Programs initiated in 2010 aimed at reduced fertilizer inputs, reduction in runoff and leaching of nutrients. It also aimed at improving water quality irrigation use efficiency.

Segarra (2001) states that PF was first tested in Northern Texas during mid 1990s. The PF was introduced as a method to economize on fertilizer based nutrient supply into fields that supported cotton, sorghum or corn. Farmers actually aimed at deriving greater profits by adopting "management zones" to place precise quantities of fertilizers and water. Surveys of cotton farms in Southern plains of United States of America showed that, PF was accepted by most farmers, mainly to lessen burden on fertilizer input, improve soil quality and most importantly gain extra cash benefits. The PF methods had spread into large portion of cotton belt by 2000 (Roberts et al., 2002).

Field experimentation with precision techniques began in early 1998 in New Mexico State. They used a series of techniques such as grid sampling, sensors to assess soil and crop nutrient status, variable rate applicators and so on. Farmers were also educated about usefulness of adopting PF. The PF was introduced in many locations within New Mexico, mainly to optimize fertilizer based nutrient supply, pesticide sprays and irrigation. Farmers in New Mexico could also obtain proper soil fertility and productivity data and revise their yield goals accordingly (Ball and Peterson, 1998).

Historically, soil sampling drew greater attention from farmers who intended intensive farming and those who wanted to replenish lost nutrients and reap uniform harvests. Soil sampling became more accurate when both surface and subsurface samples were drawn at close spacing. This provided better judgment regarding within field variations of soil fertility, especially major nutrient distribution. Intensive sampling of soil is a recent trend corresponding with adoption of PF approaches. In the North American farming zones, farmers adopting PF have consistently sampled soils using grid or management zones and analyzed them for nutrients and other traits since past 15–20 years (Ferguson and Hergert, 2009). Since 2000 A.D., remote sensed soil maps, yield maps, and maps depicting variations in soil fertility have been in vogue with farms practicing PF. During recent years, sampling densities and measurements have increased enormously. The main aim is to enhance accuracy of fertilizer supply across a field and remove unevenness in crop growth and grain harvests as effectively as possible.

According to Sparovek and Schnug (2001), historically the concept of PF emanated from North American fertilizer industry specialists, who aimed at improving soil fertility assessment and fertilizer efficiency. It was introduced into Brazilian cropping zones in South and Southeast during later part of 1990s. Argentineans have been using precision techniques, especially those related to soil maps and yield monitoring since past decade. Experimental evaluation and adoption of PF methods began in Argentina in 1996. Initially, it was tested on cereals and soybean grown in large expanses that are common to Pampas, more precisely in the Cardoba region of Argentina (Bongiovanni and Lowenberg-DeBoer, 2005). This program later expanded into other areas of the nation.

The PF techniques were in vogue in Chile by the turn of the century. Such techniques have been applied on major crops like wheat, oats, beet, and maize that are generally grown on Rhodoxeralfs (Alfisols). The PF that includes preparation of soil fertility maps; chlorophyll meter readings and variable rate N inputs have been practiced by Chilean farmers since 2000 A.D. (Claret, 2011; Molina and Ortega, 2006; Ortega et al., 2009; Villar and Ortega, 2003). Regulation of N dynamics in fields, mainly reduction in fertilizer-N supply and loss through leaching and emissions are major reasons for pursuing precision techniques.

Quest to standardize and adopt PF techniques began in many of the European nations during early part of 1990s. They adopted a range of techniques that suited different regions within Europe. Factors like topography, cropping systems, intensity of cropping, economic value of the crop, and advances in instrumentation affected the spread of PF. Sensor based technology was common in many farms that produced wheat. Fertilizer recommendations were guided by the computer based decision support and yield goals. In some regions, for example in the wheat growing regions of France, farmers started using digital imagery and remote sensing to regulate fertilizer and water supply to their fields. By the year 2002, farm cooperatives were supplied with remote sensed images and appropriate suggestions based on satellite pictures (SPOT) (Astrium, 2002).

In many parts of Southern Africa, PF is a modern technology that is fueled by recent advances in computer-based control of farm equipments and implements. Experimental evaluation and adoption of site-specific methods is more recent in South Africa. It began during early part of past decade. Much of the interest in PF in South Africa is focused around soil and fertilizer-based nutrients on crop productivity and farm profits (Maine et al., 2005). The PF techniques were also evaluated for use on horticultural and cash crops common to South Africa. Main goals were to assess its impact on soil and environment, crop productivity and economic advantages to farmers. Both short and long term benefits of PF were evaluated (Maine and Nell, 2005).

In South Africa, PF was initiated as part of a continuous 30 year steady modernization of machinery, implements and methods that improvise traction, seedings, interculture, harvest, and processing. Early efforts to introduce and standardize PF in South African farming zones occurred during mid 1990s (Rusch, 2001). Initially, techniques and instrumentation relevant to PF were derived from European nations.

Earliest of the reviews regarding adoption of PF during sugarcane production in Mauritius were made by Jhoty and Autrey (2000). They have made several suggestions regarding improvising sugarcane production through PF. They have argued that

PF approach that improves resource use efficiency (fertilizers, irrigation, and pesticides) and enhances profitability could make sugarcane production more competitive.

Based on their observation in several developing countries of Asia, Tran, and Nguyen (2008) state that PF system got initiated in early 1990s. It was practiced in various forms depending on knowledge base and available technologies. It is implementation depended on access to combination of advanced information technology and farm mechanization. Electronic methods of data collection and decision making played the key role in its adoption. The basic components of PF in any agro climatic region comprises remote sensing, GIS, GPS, soil testing, sensors, mapping of soil fertility variations, yield monitors, and VRT.

According to Indian Agricultural Research agencies, facilities for adoption of PF are available for use in many agro-ecoregions of India. The basic aspects involved are assessing soil variability, managing variation, and evaluation of impact of precision techniques. During recent years, PF has been evaluated in India, on crops like potato, rice, wheat, and cotton. It seems there are clear possibilities to economize on fertilizer-N supply and still improve crop productivity.

The PF that involves GPS, GIS, grids and soil fertility maps, sensors, VIT, and yield monitors was introduced in parts of Southern India, on an experimental basis in the dry regions of Dharmapuri in Tamil Nadu, during 2003–2004 kharif season. It included 5 crops that were managed using chisel ploughing, satellite based soil fertility maps, and fertilizer application using variable rate instruments and drip irrigation to regulate soil moisture uniformly. In three years between 2003 and 2006, number farmers adopting PF increased and the crops covered included many cereals, legumes, and vegetables. Farmers making profits from PF congregated to create "precision farming societies". Such societies disseminated knowledge among others in the area (India Development Gate, 2010; Shanwad, 2010). Reports by Maheswari et al. (2008) clearly suggest that quite a large number of vegetable farmers situated in different districts of Tamil Nadu adopted PF. In Dharmapuri District of Tamil Nadu, which is situated in dry region, economic surveys indicated that PF was in vogue by 2005 and most farmers reaped better profits compared to conventional farming procedures. During past decade (2000–2010), several other agencies initiated long term evaluation of PF. For example, Indian Space Research Organization evaluated remote sensing and soil fertility maps derived from it for potato production at Jalandhar, in Punjab. Similarly, cropping systems project at Modipuram evaluated soil fertility mapping and variable rate inputs into several crop rotations followed in Gangetic plains.

Regarding practice of PF in the horticultural belts of South India, it is interesting to note that at present, several "Precision farming Development Centers" has been started. Such centers standardize precision techniques to suit to the cropping pattern adopted locally and test them commercially in the farmer's fields (KSHMA, 2011). Most of the PF centers meant for horticultural crops are situated in the Agricultural universities and research institutes of Indian Council of Agricultural Research, New Delhi (Shanwad, 2010).

In China, effort to introduce PF was guided by the Chinese Academy of Agricultural Science, Beijing. They conducted a series of experimental evaluations in 1990s on crops like cotton, maize, and wheat in the provinces such as Hebei, Shanxi, and Shandong. Reports suggest that Chinese farmers used soil fertility maps, GIS, and GPS guided systems on maize grown in Shandong as early as 1998 (Jiyun and Cheng, 2011). State Agricultural Agencies and farmers were driven to adopt PF mainly to improve fertilizer efficiency and productivity of land per unit time.

Maize is an important cereal crop of China. It is often intercropped or rotated with soybean. Fertilizer supply and its efficient use are priority items within this cropping belt. Therefore, efforts to manage soil nutrients using PF during maize production were initiated during later half of 1990s and 2000 A.D. (Wang et al., 2006). Similarly, rice production zones are vast and intensive in many areas of China. The PF approaches were evaluated for their utility in rice production during the past decade (Wang et al., 2006; Xie et al., 2007; Zhang et al., 2006). Currently, many farmers in major cereal production zones are practicing PF.

In the Australian wheat belt, experimentation and use of PF got initiated during late 1990s. Initially, it was confined to preparing soil fertility maps and variable rate N application. At present, PF has been applied to several other aspects of wheat production (GRDC, 2010). Dobermann et al. (2004) opines that during mid 1990s, Australian farmers were enthusiastic about the newly developed yield monitors and GPS guided instrumentation. Yield monitors were introduced in the Australian wheat belt during 1993. Large farms in Western Australia used GPS guided machines for wheat production on duplex soils. However, farmers found cost of grid sampling and using yield maps prohibitive. Hence, they resorted to management zones and strip-based techniques to supply variable rates of fertilizers. During recent year that is since 2005, precision techniques based on "on-the-go" soil analysis, computer-based decision support systems and use of computer simulation/models are becoming popular.

The Australian rice belt is relatively very small, yet its productivity is high at 10 t grain ha^{-1}. The soil fertility variation that occurs during rice cultivation is being tackled using PF approaches, at least in some areas of New South Wales (Spackman et al., 2003). Precision techniques are being used to supply in season split dosages of fertilizer-N. They use both sensors placed close to crop canopy as well as multispectral remote sensing data to feed their variable rate fertilizer applicators. Reports by AGMARDT (2002) suggest that in New Zealand, precision techniques were adopted to assess soil fertility and moisture variation in fields in order to supply nutrients/irrigation using VIT on cereals and oilseeds grown as early as 1998.

1.1.1 Trends in Soil Fertility Management Practices

The PF is part of a series of soil fertility management methods that farmers and researchers have been devising and adopting since ages. Indeed several of them are still popular and useful to farmer's world over. Now, let us consider a chronology of farming techniques that were shrewdly devised to maintain soil fertility, crop productivity, and ecosystematic functions. Historically, farmers began adopting soil fertility restoration methods many centuries ago. These nutrient management techniques generally involved *in situ* measures like residue recycling, fallows, and refurbishment using animal or farmyard manures. Soil fertility was held at optimum level, to the extent possible, depending on the nutrient content of the organic manures. Variations in nutrient availability were removed only to a certain extent and it persisted. Repeated cropping and insufficient recycling meant lower grain/forage yield, occurrence of soil fertility variations within a field and in many cases nutrient deficiencies were conspicuous. As a consequence, potential grain/forage yield possible in a given location or environment was not possible. Progressively crop yield would decrease in a location.

Knowledge gained through mineral theory of crop growth, several other soil fertility concepts and fertilizer technology helped us to develop a series of different soil nutrient management procedures, all aimed at restoring soil nutrient status and maximizing crop yield. Earliest of the soil fertility management methods involved visual score of crop, identification of nutrient deficiencies and matching a supply of fertilizer-based nutrient. Although, nutrient deficiencies were overcome to a certain extent, it did not ensure optimum yield. The nutrient supply was not based on a yield goal. It only ensured removal of nutrient deficiencies transitorily. Further, nutrient ratios were also not at all maintained and this could have lead to nutrient imbalance. In many situations, it actually led to lack of yield response due to Liebig's Law of Minimum.

Later, a concept more generally applicable and based on previous evaluation of crop's response to different levels of fertilizer supply was adopted. It was called "Critical Nutrient level". Critical nutrient availability in soil is a level at which at least 95% of maximum grain/forage yield potential in the given location or environment could be produced. The critical nutrient level varies depending on several factors mostly related to soil, crop, environmental parameters, and yield goals. During past 23 decades, farmers cultivating perennial crops have been exposed to fertilizer recommendations based on plant tissue analyses. It involves elaborate sampling and analysis of all essential nutrients in leaf, leaf blade or petiole or any other portion of a plant. It is more commonly known as DRIS. Since the method utilizes only a small portion of plant tissue and just a few samples, it is non-destructive, if we consider the entire crop. Fertilizer recommendations could be altered periodically as the crop grows or seasons progress, based on nutrient status of the crop.

Next, a concept called "STCR" was envisaged. It involved series of field trials that evaluated crop's response to different levels of fertilizers. Aspects like fertilizer formulations, nutrient ratios, methods of fertilizer placement, crop genotype, and yield goals could all be standardized and recommended to farmers on a blanket basis, as suitable for a given agro-ecoregion.

Farmer's traditional practices were also in vogue in many agricultural belts. Farmers adopt several practices that maintain soil fertility and maximize crop yield. Such agronomic measures were standardized through the ages drawing knowledge from folklore and based on recent experiences. Collectively, these approaches are called

"Farmer's Traditional Practices". Farmers may apply soil test values before prescribing fertilizers. Farmer's practices envisage supply of nutrients through both inorganic and organic sources. The quantity of fertilizers supplied is guided by the soil type, crop genotype, and season. Nutrients supplied may not suffice to achieve maximum possible yield. Also, fertilizer dosages may not be the best in terms of economic advantages. Nutrient ratios in soil may or may not get satisfied. Usually application of organic manures removes dearth for micronutrients. Currently, farmer's practice in a moderately fertile belt involves inorganic fertilizers, Farmyard manure (FYM), bio-fertilizer and amendments to correct soil pH, if required.

State agency recommendation (SAR) was formulated to guide farmers in an agricultural belt or large cropping expanse. Fertilizer recommendation is primarily guided by State Agricultural Programs related to land management, cropping systems, intensity, and annual yield goal. Fertilizer supply rates are stipulated for a soil type, crop, region or yield goal. The SAR do not consider within field variations. Therefore, nutrient supply could often be higher than required or sometimes insufficient depending on variations in soil fertility. Nutrient dynamics in a large area is optimized through state agency stipulations.

Best management practice (BMP) is a term more commonly used to denote a collection of soil fertility management methods that result in high grain/forage yield and offers best economic advantages in a given location. Fertilizer supply is held at high levels so that deficiencies are not expressed. Grain yield is generally high. Soil fertility is held optimum using inorganic and organic sources. Bio-fertilizers are also used.

Maximum Yield Technology is a concept that envisages fertilizer supply so that grain/forage productivity is highest in a given locality. Yield maximization involves application of high rates of inorganic fertilizers and FYM. Basically, aspects like fertilizer quantities, their timing, ratios, and formulations are all aimed at maximizing biomass production. It does not consider soil fertility variations within a single field. Often, a certain quantity is held in subsurface layers of soil as residual nutrients.

Integrated nutrient management (INM) envisages supply of nutrients based on soil tests, crop's demand for nutrients, and yield goals. Most importantly, it considers environmental issues like soil deterioration, exhaustion of nutrients, and recycling and soil quality. Therefore, under INM, farmers are asked to supply nutrients using as many different sources. Both, organic and inorganic sources of nutrients are utilized at different ratios. In addition, bio-fertilizers are also used. Crop yields are generally optimum but not the best or maximum in a given locality. Again, INM does not consider soil fertility variations that may occur within a field. Nutrient accumulation or depletion in soil based on a given locality and cropping pattern is a clear possibility.

Site-specific nutrient management (SSNM) or PF is perhaps the most recent technique among the series of nutrient management methods that is known to farmers situated worldwide. It considers relatively minute variations in soil even within a small field. Often, it involves use of detailed grid sampling, soil nutrient estimations; preparation of soil maps, GIS, computer models and GPS guided soil fertility distribution using variable applicators. Of course, SSNM is also amenable manually. The PF is a relatively new procedure. Basically, it considers soil fertility and crop's demand for nutrients as accurately as possible in time and space. Since nutrient supplies are exact, undue accumulation in the soil profile is avoided. It also avoids soil and ground water deterioration. Crop yield is optimum and economic benefits are generally slightly more than that achieved using other procedures. Ecosystematic functions are not altered or affected to any great extent.

1.2 DEFINITIONS FOR PRECISION FARMING

According to Khosla (2011), PF has indeed experienced unprecedented expansion and popularity in some parts of the world, especially where intensive farming practices are in vogue and productivity is relatively high. The PF as a concept has been understood and utilized in different ways by the farmers and researchers. There are of course several different explanations about what is PF or not. Khosla (2011) says PF connotes several "Rs". They are right input, right timing, right amount, right place, right methods, right manner, right machinery, right crop, right fertilizers and other inputs and so on. However, he also cautions that PF is often misinterpreted as complex technological invention of the recent times, meant to be used mostly by the rich farmers and large farms. The PF is said to be costly because it involves large machinery, high electronic and mechanical sophistication, and costly labor. It is not so. The PF is amenable to both, highly mechanized farms that are controlled electronically and small farms that depend on manual distribution of seeds, farm inputs and regulation of irrigation.

The PF has been defined and explained in variety of ways depending on context, purpose or end use, relevant agricultural operations and inputs such as fertilizers, water, pesticide, methods employed and so on. Most of these definitions or explanations deal with utility of PF in managing soil fertility that is obtaining uniform and accurate distribution of soil nutrients and/or water. Following is a list of definitions and brief explanations provided by various researchers.

One of the definitions arrived at the Second International Conference on Site-specific Management for Agricultural Systems, held at Minneapolis in 1994 states that, PF or site-specific crop management is an information and technology based agricultural management system to identify, analyze and manage site-soil spatial and temporal variability within fields for optimum profitability, sustainability, and protection of the environment.

One of the most common definitions for PF states that it begins with an accurate assessment of soil fertility through GPS mapping, soil sampling, and testing. It aims to achieve maximum efficiency with regard to nutrient and water supply, and maximize profits. According to National Agricultural Research Council of United States of America, PF that is also frequently termed as "site-specific crop management" refers to developing agricultural management system that promotes variable management practices within a field. It is dependent on site's soil fertility conditions (NARC, 1997).

Berry et al. (2010) consider a wider horizon for PF. They define it as a system that focuses on sound crop production, applying geotechnology to effectively understand and manage the dynamic flows and cycles of nutrients within a field or agroecosystem.

Rickman et al. (2003) have suggested that a mismatch between uniformity of crop treatments, nutrient supply and distribution necessitates adoption of PF. The PF actually integrates a suite of technologies that retain benefits of large-scale mechanization. Yet, it recognizes local variations and aims to correct them. Satellite-based data accrual regarding soil fertility variations and crop development allows farmers to fine tune seeding, fertilizer, and water supply. It generally lowers cost of production.

Blackmore (2003) states that PF strives to improve yield goal and efficiency in agricultural practices. It involves developing techniques and procedures that help agricultural managers to enhance production efficiency. It integrates computing, electronic gadgetry, and satellite based-techniques.

Hernandez and Mulla (2008) have opined that PF is a holistic, new and developing agricultural system. It is adoption could progressively change crop production practices and trends in United States of America and other parts of the world. The PF is already influencing the agricultural crop production through adoption new GPS-based techniques. Precise spatial and temporal information regarding soil fertility and moisture can increase input efficiency, farm productivity, and profitability. At the same time, it imparts a certain influence on environmental quality of farm.

Karlen et al. (1998) have defined that precision agriculture, at the minimum, requires three elements, namely:

(a) Positioning capabilities (GPS) to know where certain equipment is located;
(b) Real-time mechanisms for controlling nutrient and water related inputs; and
(c) Databases or sensors that provide information needed to develop input schedule to suit the site-specific conditions.

Ess and Vyn (2010) define PF as an old idea provided with new life by the advent of technologies based on GPS. These GPS-based techniques are used to tailor soil and crop management in order to match conditions at every location in the field.

Precision agriculture usually includes the management of within field variability using information technology (GIS) and geopositioning methods. In addition, such new techniques help the farmer in meticulous documentation of field data related to soil fertility, water, labor inputs, and crop productivity (Vrindts et al., 2003).

Goddard (1997) had suggested that PF should not be construed as a simple yield mapping and variable rate fertilizer distribution method. Instead, it also includes production aspects like extension, management and economic advantages.

Dobermann et al. (2004) have provided a definition that considers PF as a more holistic approach with far fetching influence on agriculture. They state that, although PF started as a technology-led development, it is not just synonymous with yield mapping and VRT, for managing spatial variability within a field. Instead, PF should be considered as a systems approach to crop production, in which the goal is to reduce decision uncertainty through better understanding and management of uncontrolled variation. Expertise from many disciplines is utilized, including information technology to bring data from multiple sources and scales to bear on decisions associated with crop production.

While dealing with Malaysian rice growing areas, Gholizadeh et al. (2009) have stated that PF is a conceptualized "systems approach". It considers crop production as a total system that aims at sustainable low inputs and high efficiency. It is practically a site-specific strategy to manage production inputs and outputs.

While discussing the economic aspects, Batte and Arnholt (2003) have stated that PF has the potential to help farmers with appropriate input allocation within each field, thereby lowering quantity of fertilizers/water to be applied. In other words, it reduces on production cost and improves profitability. Batte and Van Buren (1999) suggest that precision faming or site-specific crop management is actually a combination of technologies or methods and their integration permits the following: namely:

(a) Collection of data on an appropriate scale at a suitable time;
(b) Integration and analysis of data to support a range of management options; and
(c) Implementation of a management response on an appropriate scale and at a suitable time.

Sparovek and Schnug (2001) describe PF as an umbrella terminology that encompasses knowledge (agronomic practices), and its practical expression (machines, computers, software, treatments, and procedures) in order to solve problems related to soil fertility variations.

Bramely (2006) states that precision agriculture seeks to exert greater control over a crop production system by recognizing variation and managing different areas of land differently, according to a range of economic and environmental goals.

The PF technologies are used to identify and measure within field variability and its causes. It prescribes site-specific inputs (fertilizers and/or water) that differ with crop and soil type. Reduction in input (fertilizers and/or water) levels, increased efficiency, uniformity in soil fertility or moisture, and proper timing are known to enhance cotton crop yield and cost benefit ratios (Banerjee and Martin, 2007).

Basic mapping and field level record keeping is one of the first practices under PF. The benefits of implementing PF include increased profits through increased efficiency, reduced agronomic inputs, improved production, and reduced environmental impact (Koostra et al., 2003).

The zonation within a field is an important aspect of PF. According to Aimrun et al. (2011), identification and management of spatially coherent regions (zones) is a crucial aspect under site-specific management or PF. In order to obtain maximum efficiency of inputs (fertilizers), it is said that management zones should be homogeneous combination of potential yield limiting factors.

Regarding PF practices in vogue with cotton farmers of Northern Texas; Yu et al. (2001) explain that it is a set of site-specific methods involving advanced information-based agricultural management system. It has been designed to identify, analyze, and manage spatial and temporal variability of soil effectively and obtain maximum profitability. It also aims at preserving physico-chemical characteristics and quality of soil.

The PF in European farming zone integrates GPS and GIS technologies into daily routines of farms. The PF is said to be an old traditional farming in the modern way. It involves optimizing agricultural production through improving the precision of agronomic procedures by implementing them at subfield scale.

With regard to paddy farming, Norasma et al. (2010) state that PF involves satellites, sensors, and field or thematic maps. It is a fairly comprehensive approach designed to optimize crop production by tailoring soil and crop management procedures to fit each and every field separately.

Jhoty and Autrey (2000) explain that PF is a concept based on fact that soil fertility and moisture distribution are site/location specific. Soils are feebly or immensely variable with regard to nutrient and moisture distribution, microclimate, weed species and so on. We should note that crop productivity is directly influenced by site-specific variations (Khakural et al., 1996).

According to Mishra et al. (2003), PF is a buzzword that is based on the philosophy of soil heterogeneity and homogeneity. It requires precise information on the degree of variability for fertility/water management within a field.

The PF or Precision Agriculture is a concept that involves use of new techniques, field information, adopting right agronomic practice, applying right amounts of nutrients/water, and at right time. Most importantly, information collected about specific field is utilized to evaluate and decide on optimum planting density, estimate fertilizer requirements accurately at various stages of the crop, and predict grain/forage yield (India Development Gateway, 2010). The PF avoids unnecessary or excessive use of inputs.

The PF is defined as application of principles and technologies that allow us to manage spatial and temporal variability associated with various soil fertility factors. It is relatively a comprehensive approach that optimizes crop production by utilizing data accrued *via* several sources including satellite-based information (Shylla et al., 2006).

Patil and Shanwad (2010) have defined PF keeping in view the large expanse of cereals and legumes in the Vertisol belt of South India. They state that PF aims at optimizing profitability and protecting environment through efficient use of inputs, based on temporal and spatial variability of soils and crops.

Regarding precision agriculture in the West African Sahel Florax et al. (2005) state that identification of local soil variability caused by within field differences in macronutrient availability and relevant ecological features is important, for effectiveness of PF methods. Several spatial statistical, economic, soil and crop related analyses are utilized to arrive at appropriate input levels.

In South Africa, PF is also called as Computer-aided Farming systems. It is defined as a process whereby, a large field is divided into finite number of subfields, allowing variation of inputs in accordance with the data gathered. It results in maximization of profits, minimizes risks like nutrient accumulation and reduces ill effects on agro-environment (Rusch, 2001). Godwin et al. (2001; 2003) described PF as a name given to a method of crop management that entails management of areas within a crop field that require different levels of input.

The PF or "Prescription farming" or "Site-specific Crop Management" involves collection of site-specific data in order to make appropriate decisions, regarding nutrients, and water supply to discrete areas within a field. The PF utilizes GPS, GIS, and VRT (Ball and Peterson, 1998).

Roberts et al. (2002) state that, PF of cotton in Tennessee involves use of set of technologies to identify and measure within field variability and its causes. It then prescribes site-specific inputs (fertilizer, water) that match varying crop and soil needs during a crop season.

Regarding fertilizer-N inputs to cotton crop cultivated in Tennessee, Torbet et al. (2008) state that PF improves accuracy of fertilizer-N supply by using data and information drawn through various GIS and GPS methods. Geospatial mapping, crop development, and yield sensing help in improving fertilizer-N efficiency.

Definitions that relate to role of PF in improving management of within field soil N and moisture variations are available in plenty. According to Shaver et al. (2010), one of the primary goals of PF is to regulate on-farm inputs such as fertilizer-N by determining in-field variability that directly affects crop growth and development. Data derived from several methods could be used to decide most appropriate quantity, timing and distribution methods.

Al-Kufaishi (2005) explains that with regard to irrigation, primary goal of PF is to supply the growing crop with right amount of water at the right time and in the right place, avoiding any surplus that could lead to the leaching of water and nutrients that limits plant growth and reduce yield.

1.3 MAJOR COMPONENTS OF PRECISION FARMING OR SITE-SPECIFIC FARMING

The PF or site-specific farming is a set of geospatial technologies and accurate procedures that actually link mapped locations in a field with most appropriate decisions regarding seeding, fertilizer inputs, irrigation, plant protection chemicals, and yield recording (Berry et al., 2010; Davis et al., 1998). The PF in any agricultural zone has following components at the least. They are:

(a) GPS,
(b) Remote Sensing Imagery or Sensor data regarding soil fertility variations and crop productivity,
(c) GIS software, and
(d) Variable Rate Applicators (VRT with robotics that are guided by computer models dealing with soil fertility status and crop response data.

1.3.1 Precision Farming: A Map and/or Sensor based Technology

Morgan (1997) has suggested that at least two sets of methodologies could be used during PF. One of them is termed map-based and the other sensor-based. Map-based methods involve grid sampling, laboratory analysis of soil samples at least at two depths, generating soil fertility and nutrient distribution maps or maps that depict soil physico-chemical characteristics like pH, cation exchange capacity (CEC), electrical conductivity, and so on. The map generated using soil data is then used to guide a variable applicator. During both soil sampling and variable rate inputs, we need a positioning system—Differential Global Positioning System (DGPS) to identify each location. The second method, usually called Sensor-based PF involves extensive use of real-time sensors and feedback control system to measure soil properties rapidly "on-the-go". Then, immediately use these signals to direct a variable rate applicator appropriately. Often tractors are fitted with sensor in the front or on a separate vehicle that leads the tractor fitted with variable applicator. This allows sufficient time for data accrual *via* sensors, then to process it using computer models and direct variable applicators. The sensor-based technique does not need a GPS system, if treatments are made immediately. However, traction equipment fitted with GPS can be loaded and data preserved for posterity. Let us consider an example. "Soil DoctorR" produced and marketed by Crop Technology Inc, Houston, Texas, USA, is an example that examines, soil type, organic matter content, CEC, soil moisture, and NO_3-N using a rolling electrode as the tractor moves through the field. The need for GPS is eliminated. Yet another example is a Sensor-based technology developed by researchers at Purdue University, Indiana, USA. It is a sensor (photodiode) surrounded with Light Emitting Diode (LED). The reflected light is measured to estimate soil organic matter. Soil moisture can also be estimated. Examples pertaining to real time sensors that estimate soil texture, pH, NO_3-N, K, and P are available in literature.

1.3.2 Global Information Systems (GIS)

In general, GIS enhances our ability with regard to decision support system and planning of fertilizer, water and pesticide distribution. It is actually organized collection of computers that help in data capture, storage, retrieval, update, manipulation, and analysis. It also helps us in integrating geographical data with various aspects like field history, soil fertility, crop husbandry, and crop production projections. It allows us to simulate fertilizer supply and crop response in a location, so that appropriate decisions could be made. The GIS relevant to crop production allows us to analyze data collected through a period of time at a given location. GIS allows us to study spatial effects of soil management and agronomic factors on crop productivity. For example, effects of soil type, its texture or fertility status, pH, organic matter or moisture on grain/forage yield could be ascertained using spatial data. Actually, regression analysis of data regarding influence of major nutrients (N, P, and K) from entire field could be stored and used to derive appropriate decisions in the following season. Soil nutrient maps, maps that depict spatial variations pertaining to several other relevant soil characters like texture, pH, electrical conductivity, and so on could be effectively used. GIS helps in preparing prescription maps for each field or smaller management zones based on spatial data and crop production trends (Berry et al., 2010).

In addition to information on soil fertility and nutrient distribution within specific fields, GIS records for larger cropping zones and stretches that encompass a state or a small nation are also available. They help us in channeling fertilizers and prescribing cropping patterns more accurately in a given cropping belt. For example, such detailed information on soil nutrient status and pH variations in different states and sub-regions are available in USA and many European nations. Fertilizer recommendations that are arrived after considering a range of natural factors, crop production systems, yield goals, and most importantly the economic advantages are also available to be retrieved and consulted quickly using GIS (Bundy et al., 2005; Laboski and Bundy, 2005). A recent report suggests that NLEAPGIS are a tool that helps farmers with US soil database and climatic parameters. It aids in soil N management and risk assessment (Delgado, 2011). This facility covers entire USA. Similarly, soil nutrient distribution and productivity data for countries in Latin America, Africa and Asia are available for retrieval through computers.

Web-based GIS Decision support systems are also available in certain regions. Its functions are comparable to other computer-aided services like e-commerce, information sharing and disseminating web sites. It is an open source technology that allows free access anywhere in the world. For example, researchers at Universiti Putra, in Malaysia have explored the use of Minnesota open map server, hypertext preprocessor, Apache Web server and MySQL database to study the soil fertility variations in paddy growing regions (Norasma et al., 2010). It actually allows the farmers to access webs that carry information on rice cultivation procedures and economic advantages due to them.

Historical data on paddy planting dates, fertilizer input schedules, agronomic practices, growth and yield pattern, and yield response data could be retrieved or used directly to arrive at computer-aided decisions. This system also allows integration with soil fertility maps and variable rate inputs of fertilizers-N, P, and K. Farmers may of course print soil maps or carry digital information separately and feed it to their vehicles carrying variable rate applicators. It is believed that such web-based decision support will become fairly common across farming zones in all continents. It is adoption involves least cost to the farmers. Yet, it allows farmers with large and wide array of information on which they could base their agronomic procedures. It also allows rapid sharing of information. It should be a helpful tool to policy makers involved in deciding cropping pattern, input supply, and yield goals.

1.3.3 Global Position Systems (GPS)

The GPS are techniques that provide farmers with unequalled accuracy, flexibility in positioning farmer's vehicles (tractors), navigation, and data capture about a particular spot, a field or farm. The spatial variations are easily deciphered and stored through GPS satellites. The GPS technology is known to use a set of 2431 satellites, situated in high altitude orbit above earth. They are focused to survey earths' surface soil, water, and crops. These satellites transmit signals continuously that are picked up by special receivers. The GPS is divided into three aspects:

(a) A space segment that includes 21 operational satellites that orbit the earth at 20,000 km,

(b) A control system in Colorado, in central USA and linked stations, and

(c) A user segment that consists of receivers which help in positioning, velocity and time information to the user. A GPS receiver, it is said requires at the least contact with four satellites to set its coordinates and location on earth. The GPS signals have to be sharp and accurate. Raw and hazy signals are not useful to determine a position in a field. The GPS referenced signals are often used to plant seeds prepare soil fertility maps, yield variations, soil moisture distribution and so on (Plate 1).

PLATE 1 Planting Maize in Iowa, USA.

Source: Mr David Nelson, Nelson Farms Inc, Fort Dodge, USA

Note: Planting Maize and Soybean using GPS guided seeders is common in the Northern plains.

A medium range GPS receiver can establish positions within a field with an accuracy of 1.0 m between two spots (Berry et al., 2010). Such receivers allow us to stamp data pertaining to soil nutrient status, moisture or pH and so on with accurate geographic coordinates to locate them at any time. Currently, combination of two or more constellations of satellites, like US GPS system and Russian GLONASS has improved accuracies to 2 cm. The GPS aided field boundaries and management zone demarcation are becoming common. In areas prone to soil maladies like salinity, GPS can be coupled with salinity meter sledge and towed on a pick up van, all along the field. Salinity mapping is an important procedure in the areas afflicted with high salts. The GPS receivers are also used to map weed intensity and spread in a field. It helps while accurately distributing herbicides in a field. The DGPS uses two GPS receivers–a base station that is stationary and located at a referenced point and the other is mobile fitted on the vehicle. The DGPS helps in overcoming problems related to resolution.

Incidentally, Franzen et al. (2008) have cautioned about use of the word "GPS" in place of PF. The GPS actually makes use of a series of military satellites to determine a precise geographic location. Deciphering exact location of farm vehicles or a point in a field accurately has its advantages. The GPS helps in geographic identification of soil properties. The GPS helps in distributing fertilizers accurately at each point based on soil maps. The GPS helps in accurately mapping crop growth productivity through yield monitors. The GPS helps in identifying areas that support low or high yield.

A satellite system equivalent to GPS is available to farmers in Europe. It was developed by former Soviet Union and is called GLONASS. It is controlled by CIS space command. It is a satellite system that has about 24 satellites (Blackmore, 2003).

Variable Rate Technology

The VIT involves application of seeds, fertilizers, and irrigation or pesticides in quantities that is specific to each spot and variable (Plate 1). Inputs are decided by field map and a decision support system. For example, in case of soil fertility, a computer-based decision support considers soil fertility maps, nutrient availability pattern, crop species/genotype, yield goals, and profitability. The quantity of input such as fertilizers or water channeled at each spot depends on directions received from decision support systems by a GPS guided vehicle. Variable rate fertilizer supply provides uniformity with regard to soil fertility in an individual field. Similarly, variable rate irrigation based on soil moisture map creates uniform availability of soil moisture to the crop. Details on VIT and instrumentation involved are made available in Chapter 2.

1.4 ADVANTAGES THAT ACCRUE DUE TO PRECISION FARMING

The PF offers a wide range of advantages to farmers. The extent of advantages derived is based on geographic location and farming enterprise in question. Yet, we can generalize and group the advantages. For example, Tran and Nguyen (2008) have suggested that extent and range of benefits accrued due to adoption of PF may differ between developing and developed nations. Obviously, low or moderate supply of fertilizer-based nutrients, water and other farm amendments may offer commensurate gains in developing nations. Whereas, high input trends in agriculturally developed regions may allow us proportionately higher advantages in terms of soil quality, its productivity, economic gains, and environment related advantages.

With regard to rice farming in South Asia and Fareast, Segarra (2002) have grouped the advantages from PF as follows:

- *Overall Grain/Forage Yield*: Aspects like, precise genotype, exact fertilizer input in as many splits as possible, proper nutrient ratios, appropriate irrigation schedules and timely monitoring of growth/maturation results in greater grain/forage yield, compared with similar levels of uniform or blanket application of fertilizers and irrigation.
- *Improved Efficiency of Inputs*: Aspects like, use of advanced electronically controlled machinery, geospatial techniques of identification of soil fertility/moisture variations, advanced computer-based growth and grain yield models that allow us to distribute variable and exact quantities of nutrients/water in space and time, contribute to enhanced efficiency of inputs.
- *Reduced Production Costs*: For a similar grain/forage yield level, fertilizer-based nutrient supply and irrigation required is usually marginally or conspicuously smaller. It results in lessening of input costs to farmers. Generally cost per unit grain/forage yield is lower in farms maintained using precision techniques.
- *Better Decision Making*: Agricultural machinery, satellite guided geospatial techniques and computer models allow us to accrue data and therefore helps farmers in taking decisions regarding farm operations especially nutrients and water supply.
- *Reduced Environmental Impact and Risks*: Placement of nutrients and water in accurate quantities at appropriate timings during crop growth results in rapid removal of nutrients. It reduces undue accumulation of fertilizer-based nutrients in the soil profile or ground water. Loss of nutrients like N to ambient atmosphere is lessened.
- *Accrual of Accurate and Easily Retrievable Data about a Farm or Region*: Actually almost all farm operations; their intensity and timing are recorded. Data on exact quantities of nutrients supplied to each fertility zone is available, since soil fertility maps and variable applicators are computer controlled with data retrieval facility. It definitely allows management of farm operations as we move from one season or crop to next in sequence in the same field.

We should note that advantages that accrue from PF are also specific to geographic area, soil, cropping pattern, agronomic requirements of the crop, and economic value of the produce.

For example, in the European Plains, Cragg (2004) suggests that major advantages are as follows:

(a) Precision technique saves on costs. It allows us to apply fertilizer-P and K at rate to match crop's need.
(b) The PF improves efficiency of seeding and fertilizer supply to fields.
(c) The PF improves the accuracy of crop husbandry by providing soil map, yield maps, and accurate placement of nutrients.
(d) The PF offers environmental benefits by reducing on fertilizer and chemical supply to fields.

1.5 CONSTRAINTS TO ADOPTION OF PRECISION FARMING

Precision techniques are not easily amenable to all agricultural conditions. Farmer's stipulations and desires in each continent and its sub-region too vary enormously. Constraints faced during adoption and standardization of precision techniques is indeed many.

Bongiovanni and Lowenberg-DeBoer (2005) have stated that major constraints to adoption of PF in Argentina and other Latin American countries are high initial investments on equipments and management time (labor), lack of significant variations in the soil fertility, tendency to use low amounts or no fertilizers during cereal production, risk in grain pricing, and profits from precision technology. Above all, in several cases, precision techniques fetch only small increases in returns due to low value of the product.

Cragg (2004) opines that following are the major constraints to adoption of PF in the European farming regions.

They are:

(a) Procedures to be adopted and instrumentation are too elaborate and time consuming. At times procedures get complicated.

(b) The PF needs initial investment that could be prohibitive.

(c) Interpretation of data drawn from digital imagery and other procedures could involve specific skills and may add up to already high costs.

(d) Precision involves data collection that could be costly. The remote sensing and digital imagery of fields showing variations in organic matter, nutrients and water are generally costly.

(e) The PF may not turn out to be immediately profitable. Sometimes its benefits are perceived, only if adopted on a long run for example environmental benefits. Also, often the reduction in fertilizer supply that occurs due to PF may be marginal, not perceived significantly, if the farm is small and value of the crop is low.

According to Srinivasan (2010) constraints to adoption and popularization of PF in the Asian cropping zones are as follows:

(a) High cost of obtaining site-specific data;

(b) Lack of willingness to share spatial data among various organizations;

(c) Complexity of tools and techniques that require new skills;

(d) Culture, attitude and perceptions of farmers including resistance to adoption of new techniques and lack of awareness of agro-environmental problems;

(e) Small farms, heterogeneity of cropping systems, and land tenure/ownership restrictions;

(f) Infrastructure and institutional constraints including market imperfections;

(g) Lack of success stories about adoption of PF and lack of demonstrated impacts on yields;

(h) Lack of local technical expertise;

(i) Uncertainty on returns from investments to be made on new equipment and information management systems;

(j) Inadequate understanding of agronomic factors and their interaction;

(k) Lack of understanding of the geostatistics necessary for displaying spatial variability of crops and soils using current mapping software; and

(l) Limited ability to integrate information from diverse sources with varying resolutions and intensities.

In general, researchers have consistently strived to remove lacunae and improvise precision techniques, so that it matches farmer's economic requirements and at same time answers environmental concerns.

KEYWORDS

- **Geographic information system**
- **Global positioning systems**
- **Precision farming**
- **Soil fertility**
- **Variable rate technology**

REFERENCES

AGMART. *Optimizing Yield and Profit with Precision Farming: Canterbury Precision Farming*, pp. 1–40 (2002), Retrieved from http://www.nutrientsolutions.co.nz/pdf/precision_Agriculture-initial_Nzarable_studies.pd.f (July 2nd, 2011).

Aimrun, W., Amin, M. S. M., and Amin H. Nouri. Paddy field zone characterization using Apparent Electrical Conductivity for Rice Precision Farming. *International Journal of Agricultural Research*, **6**, 10–28 (2011).

Al-Kufaishi, S. A. *Precision Irrigation*. Master's Thesis, The Royal Veterinary and Agricultural University, Denmark, p. 23 (2005).

Astrium. *Farmstar: Crop management with SPOT*, pp. 1–3 (2002), Retrieved from http://www.spotasia.com.sg/ web/sg/2625-precision-farming.php (April 25th, 2011).

Ball, S. T. and Peterson, J. L. *Precision farming in New Mexico: Enhancing the Economic Health of Agriculture*, p. 8 (1998), Retrieved from http://www.cahe.nmsu.edu/pubs/_z/z-106.html (January 6th, 2011).

Banerjee, S. and Martin, S. W. *Summary of Precision-farming practices and perceptions of Mississippi Cotton Producers*. Mississippi Agricultural and Forestry Experiment Station Bulletin No 1157, pp. 1–43 (2007).

Batte, M. T. and Arnholt, M. W. Precision farming adoption and use in Ohio: Case studies of six leading-edge adopters. *Computers and Electronics in Agriculture*, **38**, 125–139 (2003).

Batte, M. T. and Van Buren, R. N. *Precision farming: A factor influencing Productivity*. Paper presented at the Northern Ohio Crops Day Meeting, Woody County, Ohio, USA, p. 21 (1999).

Berry, J. K., Delgado, J. A., and Khosla, R. Precision Faming advances Agricultural sustainability (2010), Retrieved from www.ars.usda.gov/research/publication/publication.html?seq_no_211567 (May 28th, 2011).

Blackmore, S. *The Role of Yield Maps in Precision Farming*. National Soil Resources Institute, Cranefield University, Silsoe, United Kingdom (Doctoral Dissertation), p. 161 (2003).

Bongiovanni, R. and Lowenberg-DeBoer, J. *Precision Agriculture in Argentina*. Third Simposio Internacional de Agricultura de Precision, pp. 1–14 (2005), Retrieved from http://www.cnpms. embrapa.br/siap2005/palestras/SIAP3-Palestra_Bongiovanni_e_LDB.pdf (March 25th, 2011).

Bramely, R. *Precision Agriculture: Profiting from Variation*. Ecosystems Sciences Division, Council of Scientific and Industrial Research Organization, Glen Osmond, Australia, pp. 1–3 (2006), Retrieved from http://www.csiro.au/science/PrecisionAgriculture.htm (January 28th, 2010).

Bundy, L., Andraski, T., Laboski, C., and Sturgul, S. Determining optimum nitrogen Application rates for Corn. Department of Soil Science, University of Wisconsin-Madison, USA. *New Horizons in Soil Science*, **1**, 1–11 (2005).

Claret, M. M., Urrutia, R. P., Ortega, R. B., Stanely, B. S., and Valderrama, V. N. Quantifying Nitrate leaching in irrigated Wheat with different Nitrogen fertilization strategies in an Alfisol. *Chilean Journal of Agricultural Science*, **71**, 148–156 (2011).

Cragg, A. *The promise, pitfalls, and practicalities of Precision Farming*. HGCA Conference on Managing Soil and Roots for Profitable Production, pp. 1–6 (2004), Retrieved from www.hgca.com/events.past_events/…/press%20releases.mspx.htm (May 28th, 2011).

Davis, G., Casady, W., and Massey, R. Precision Agriculture: An Introduction. University of Missouri Extension Services. *Water Quality*, **450**, 1–9 (1998).

Delgado, J. A. A new GIS approach to assess Nitrogen Management across the USA, pp. 1–8 (2011), Retrieved from http://www.icpaonline.org/finalpdf/abstract_176.pdf (June 21, 2011).

Dobermann, A., Blackmore, S., Cook, S. E., and Adamchuk, V. I. Precision Farming: Challenges and Future Directions. In *New Directions for a Diverse Planet*. Proceedings of the Fourth International Crop Science Congress, Brisbane, Australia, p. 19 (2004), Retrieved from www.cropscience.org.au (January 20th, 2011).

Earl, R., Wheeler, P. N., Blackmore, B. S., and Godwin, R. J. Precision farming: The Management of Variability. *Landwards*, **51**, 18–23 (1996).

Ess, D., Vyn, T., and Erickson, D. *Site Specific Management*, p. 2 (2010), Retrieved from http://www. agriculture. purdue.edu/ssmc/Frames/main.html (December 15th, 2010).

Fairchild, D. Precision faming concepts an industry's perspective and experience since 1986 In *Proceedings of Site-specific Management for Agricultural Systems*. P. C. Roberts (Ed.) American Society of Agronomy, Wisconsin, USA, pp. 753–755 (1994).

Ferguson, R. W. and Hergert, G. W. Soil sampling for Precision Agriculture. *Journal of Animal and Plant Sciences*, **5**, 494–506 (2009).

Franzen, D., Hofman, V., and Halverson, A. Soil fertility and Site-specific farming study at area IV Research Farm, pp. 1–3 (2008), Retrieved from http://www.soilsci.ndsu.nodak.edu/Franzen/scientific-pubs.html (May 28th, 2011).

Gerhards, S., Wyse-Pester, D. Y., and Mortensen, D. A. Spatial stability of Weed patches in Agricultural fields. In *Proceedings of the 3rd International Conference on Precision Agriculture*. Minneapolis, Minnesota, USA, pp. 521–529 (1996).

Gholizadeh, A., Amin, M. S. M., Anuar, A. R., and Aimrun, W. Evaluation of SPAD-Chlorophyll meter in two different Rice growth stages and its temporal variability. *European Journal of Scientific Research*, **37**, 591–598 (2009).

Goddard, T. *What is Precision Farming*. Proceedings of Precision Farming Conference, Taber, Alberta, Canada, pp. 1–5 (1997).

Godwin, R. J., Earl, R., Taylor, J. C., Wood, G. A., Bradley, R. I., Welsh, J. P., Richards, T., Blackmore, B. S., Carver, M. J., Knight, S., and Welti, B. *Precision Farming of Cereal crops*. A five-year Experiment to develop Management guidelines. Home grown Cereals Authority Project no 267, p. 28 (2001).

Godwin, R. J., Richards, T. E., Wood, G. A., Welsh, J. P., and Knight, S. M. An Economic analysis of the potential for Precision farming. *Biosystems Engineering*, **84**, 533–545 (2003).

GRDC. *Precision Agriculture-Fact sheet-How to put Precision Agriculture into practice*. Grain Research and Development Corporation, Kingston, Australia, pp. 1–6 (2010), Retrieved from www.grdc.com.au (January 1st, 2011).

Griffin, T. W., Popp, J. S., and Buland, D. V. Economics of Variable rate applications of Phosphorus on a Rice and Soybean rotation in Arkansas. *Proceedings of the 5th International Conference on Precision Agriculture and other Resource Management*. Bloomington, Minnesota, USA, pp. 23–40 (2000).

Hernandez, J. A. and Mulla, D. J. *Introduction to Precision Agriculture*, pp. 1–7 (2008), Retrieved from http:// soils.umn. edu/academics/classes/soil4111.htm (December 15th, 2010).

India Development Gateway. *Precision Farming*. Tamil Nadu Agricultural University, Coimbatore, India, pp. 1–3 (2010), Retrieved from http://www.tnau.ac.in/horcbe/hitechfld.swf. (December 15th, 2010).

Jhoty, I. and Autrey, J. C. *Precision Agriculture-perspectives for the Mauritian Sugar Industry*. Mauritius Sugar Industry Research Institute, Mauritius, Bulletin 12, pp. 1–7 (2000).

Jiyun, J. and Cheng, J. Site Specific nutrient Management in China: IPNI-China program. International Plant Nutrition Institute, Norcross, Georgia, USA, pp. 1–7 (2011), Retrieved from http://www.ipni.net/ppiweb/china.nsf/$webindex/27D05B2887D6B7EE482573AE0029344 8?opendocument.

Karlen, D. L. Andrews, S., Colvin, T. S., Jaynes, D. B., and Berry, E. C. Spatial and temporal variability in corn growth, development and yield. In *Proceedings of 4th International Conference on Precision Agriculture*. American Society of Agronomy, Madison, Wisconsin, USA, pp. 101–112 (1998).

Kessler, M. C. and Lowenberg-DeBoer, J. Regression analysis of Yield monitor data and its use in fine-tuning crop decisions. *Proceedings of the 4th International Conference on Precision Agriculture*. American Society of Agronomy, Madison, WI, USA, pp. 821–828 (1998).

Khakural, B. R., Robert, P. C. and Mulla, D. T. Relating Corn and/or Soybean yield to variability in Soil and Landscape characteristics. In *Proceedings of the 3rd International Conference on Precision Agriculture*. P. C. Roberts (Ed.). American Society of Agronomy, Minneapolis, Minnesota, USA, pp. 117–128 (1996).

Khanna, M., Epouche, O. F., and Hornbaker, R. Site-specific Crop Management: Adoption patterns and incentives. *Review of Agricultural Economics*, **21**, 455–472 (1999).

Khosla, R. *Precision Agricultural Techniques and Technologies Presidential Address*. First Indo-US Bilateral workshop on Precision Farming. Punjab Agricultural University, Ludhiana, India, p. 1 (2011), Retrieved from www.agsci.colostate.edu.news/News2011/khosla_India.html (May 28th, 2011).

Koostra, B. K., Stombaugh, T. S., and Dowdy, T. C. Development of Precision Agriculture to suite using ArcPad. ESRI Paper No. 0477, (2003), Retrieved from www.bae.uky.edu/~tstomb/StombaughPackage_8-26-05.pdf (December 15th, 2010).

KSHMA. *Precision Farming Development and Extension through PFDCs*, pp. 1–17 (2011), Retrieved from http://horticulure.kar.nic.in/dshma/guidlines.html (March, 2011).

Laboski, C. and Bundy, L. New Nitrogen Rate guidelines for Corn in Wisconsin using a regional approach. Department of Soil Science, University of Wisonsin-Madison. *New Horizons in Soil Science*, **3**, 1–8 (2005).

Lowenberg-Deboer, J. Precision Framing or Convenience Farming, pp. 1–32 (2003a), Retrieved from http://www. regional.org.au/au/asa/2003/i/6/lowenberg.htm (March 23rd, 2011).

Maheswari, R. Ashok, K. R., and Prahadeeswaran, M. Precision Farming Technology, Adoption Decisions and Productivity of Vegetables in Resource-poor Environments. *Agricultural Economics Research Review*, **21**, 415–424 (2008).

Maine, N. and Nell, W. T. *Strategic approach to the implementation of Precision Agriculture principles in Cash crop farming*, pp. 217–225 (2005), Retrieved from http://www.farming-success.com /id126.htm (February 21st, 2011).

Maine, N., Nell, W. T., Alemu, Z. G., and Barker, C. Economic analysis of Nitrogen and Phosphorus application under Variable and Whole field strategies in the Bothaville district of South Africa. *Research Report of Department of Geography*. University of Free State, Bloemfontein, South Africa, p. 10 (2005), Retrieved from http://ideas.repec.org/a/ags/agreko/7047.html.

Mishra, A., Sundermoorthi, K., Chidambara Raj, and Balaji, D. *Operationalization of Precision Farming in India*. Map India Conference. New Delhi, India, (2003), Retrieved from gisdevelopment.net/ application/agriculture/.../pdf/127.pdf (December 15, 2010).

Molina, M. and Ortega, R. Evaluation of the nitrification inhibitor DMPP in two Chilean soils. *Chilean Journal of Plant Nutrition*, **29**, 521–534 (2006).

Morgan, M. T. Precision Farming: Sensors and Map-based. *Agricultural and Biological Engineering Reports*. Purdue Research Foundation, West Lafayette, Indiana, USA, pp. 1–3 (1997) http:/www.agriculture.purdue.edu/ssmc/.../sensors.html (May 28th, 2011).

NARC. *Precision Agriculture in the 21st Century: Geospatial and Information Technologies in Crop Management*. National Academy Press, Washington. DC, p. 149 (1997).

Norasma, C. Y. N., Shariff, A. R. M., Amin, A. S., Khairunniza-Bejo, S., and Mahmud, A. R. Web-based GIS Decision Support System for Paddy Precision farming. *Web Precision Farmer V.2.0*. University Putra Malaysia, Selangor, Malaysia, pp. 1–3 (November 29th, 2010).

Ortega, R. A., Munoz, R. E., Acosta, L. E., and Riquardo, J. S. Optimization model for Variable rate application in extensive crops in Chile: The effects of fertilizer distribution within the field. In *Proceedings of the 7th EFITA Conference, Wageningen*. European Federation for Information Technology and Agriculture (EFITA), Wageningen. Netherlands, pp. 489–495 (2009).

Patil, V. C. and Shanwad, U. K. *Relevance of Precision Farming to Indian Agriculture*. Department of Agronomy, University of Agricultural Sciences, at Raichur, Karnataka, India, (2010), Retrieved from www.acr.edu.in/info/infofile/144.pdf (May 28th, 2011).

Rickman, D., Luvall, J. C., Shaw, J., Mask, P., Kissel, D., and Sullivan, D. *Precision Agriculture: Changing the Face of Farming*. American Geological Institute, Alexandria, VA, USA. Geotimes, pp. 1–9 (2003), Retrieved from http://www.agiweb.org/geotimes/nov03/feature_agric.html #author (January 4th, 2011).

Roberts, R. K., English, B. C., Larson, J. A., Cochran, R. L., Goodman, B., Larkin, S., Marra, M., Martin, S., Reeves, J., and Shurley, D. *Precision farming by Cotton producers in six Southern states: Results from 2001 Southern Precision farming survey*. Department of Agricultural Economics, University of Tennessee, Knoxville, Tennessee, USA Research Series 03–02, pp. 1–83 (2002).

Rusch, P. C. *Precision Farming in South Africa*. MSc Thesis, Department of Agriculture and Food Engineering, University of Pretoria, South Africa, p. 51 (2001).

Shanwad, U. K. *Precision farming: Dreams and Realities for Indian Agriculture*, pp. 1–4 (2010), Retrieved from www.gisdevelopment.net/application/agriculture/overview/mi04115pf.htm (January 15th, 2011).

Shaver, T., McCuen, R. H., Ferguson, R., and Shanahan, J. *Crop canopy Sensor utilization for Nitrogen management in Corn under semi-arid limited irrigation conditions*. American Society of Agronomy Annual International Meetings, Long Beach California, USA (Abstract), p. 1 (2010).

Shylla, B., Handa, A., and Sharma, U. *Precision Farming in Horticulture*. Science Tech: Entrepreneur, pp. 1–5 (2006), Retrieved from http://www.techno-preneur.net/information-desk/sciencetech-magazine/2006/jan06/Precision-Farming.pdf (May 28th, 2011).

Spackman, S., Lamb, D., Louis, J., and Mackenzie, G. *Remote Sensing as a potential Precision Farming Technique for the Australian Rice Industry*, pp. 1–3 (2003), Retrieved from http://www.regional.org.au/au/gia/13/403spackman.htm (January 11th, 2011).

Sparovek, G. and Schnug, A. Soil tillage and Precision Agriculture: A theoretical case study of soil erosion in Brazilian Sugarcane production. *Soil and Tillage*, **61**, 47–54 (2001).

Srinivasan, A. *Relevance of Precision Farming Technologies in Asia and the Pacific*, pp. 1–14 (2010), Retrieved from http://www.sristi.org/mtsa/s_7-4.htm (30th July, 2011).

Sudduth, K. Current Status and Future Directions of Precision Agriculture in the USA. In *Proceedings of 2nd Asian Conference on Precision Agriculture*. Pyeongtaek, Korea, CDROM (March 1st, 2011) (2007).

Torbett, J. C., Roberts, R. K., Larson, J. A., and English, B. C. Perceived improvements in Nitrogen Fertilizer efficiency from Cotton Precision Farming. *Computers and Electronics in Agriculture*, **64**, 140–148 (2008).

Tran, D. V. and Nguyen, N. V. The concept and implementation of Precision Farming and Rice Integrated Crop Management systems for sustainable production in the twenty first century. *A report on Integrated Systems*. Food and Agricultural Research Organization of the United Nations, Rome, Italy, pp. 91–101 (2008).

Villar, D. and Ortega, Y. R. Medidor de Clorofila. Bases teoricas su application para la fertilizacion nitrogenada en cultivos. *Agronomia y Forestal UC*, **5**, 4–8 (2003).

Vrindts, E., Reyniers, M, Darius, P., Frankinet, M., Hanquet, B., Destain, M., and Baerdemaeker, J. Analysis of spatial soil, crop and yield data in a winter wheat field. *Society for Engineering in Agricultural, Food and Biological Systems*. Proceedings of the Annual International Meeting. Paper No. 031080, pp. 1–14 (2003).

Wang, H., Jin, J., and Wang, B. Improvement of Soil Nutrient Management *via* information technology. *Better Crops*, **90**, 30–32 (2006).

Xie, G., Chen, S., Qi, W., Lu, Y., Yang, X., and Liu, C. A multidisciplinary and Integrated study of Rice Precision Farming. *Chinese Geographical Science*, **13**, 9–14 (2007).

Yu, M., Segarra, E., Watson, S., Li, H., and Lascano, R. J. Precision Farming Practices in irrigated Cotton production in the Texas High plains. *Proceedings of the Beltwide Cotton Conference*, **1**, 201–208 (2001).

Zhang, J. H., Wang, K. Bailey, J. S., and Wang, R. C. Predicting Nitrogen status of Rice using Multi-spectral Data at Canopy scale. *Pedosphere*, **16**, 108–117 (2006).

2 Precision Farming: Methodology

CONTENTS

2.1 INTRODUCTION

Within the context of this book, crop's response to soil heterogeneity is the crux of the problem that precision farming intends to solve as efficiently as possible. Spatial variations in soil may occur at different scales, such as between agro-ecoregions or vast expanses of cropping zones, between farms, fields or even within a field or strip. A wide range of soil characteristics may be responsible for heterogeneity. Soil chemical properties like nutrients, their physico-chemical nature, pH, salinity, organic matter, liming pattern, and any other amendment regularly added to soil might all contribute to soil heterogeneity. Continuous cropping and variable exhaustion of nutrients is one of the important reasons for soil heterogeneity that perpetuates, if uncorrected. Spatial variations in soil nutrients may also occur due to uneven leaching, erosion, seepage, and emission rates. Soil physical properties that vary and contribute soil heterogeneity are soil texture, structure, depth, aeration, and porosity. The crop growth and grain yield response varies within a field/region mainly due to the factors and extent or type of agronomic measures adopted. Precision farming or variable rate nutrient supply is among the latest techniques that is based on satellite guided or handheld instruments. Such instruments detect soil heterogeneity or variations in nutrient availability rapidly and respond accurately based on yield goals fixed.

There are indeed several different parameters that could be measured to arrive at an understanding about soil heterogeneity and crop's response to it. Often, crop biomass distribution in terms of forage and/or grain yield is assessed. Combine harvesters or harvesting equipments fitted with GPS are used to rapidly note and prepare yield maps highlighting the variations. Harvesters with specific instrumentation to assess variability of cereal grain/forage yield or potato tubers or tomato fruits or cotton lint are available. Multispectral analysis and images from satellites could also be used to obtain information regarding variations in crop growth, nutrient status, and forage production.

There are different approaches to study soil heterogeneity (Hellebrand and Umeda, 2004). Soil samples from fields could be analyzed directly and maps that depict variations in soil mechanical and physico-chemical properties could be prepared. Soil penetrometers are useful in preparing maps about soil mechanical properties. Indirectly, the soil resistance to penetrations gives an indication about variations in soil texture, density, and moisture. Non-contact methods that allow us to assess soil physical parameters are based on Electromagnetic Induction (EMI). Field scale maps of electrical conductivity (EC) variations can be prepared using GPS and data loggers. Soil moisture variations could be assessed using Nuclear Magnetic Resonance (NMR). The proton-NMR technique allows us to measure soil moisture up to 15 cm depth in the profile. Sensing and mapping variations in soil nutrient content/availability in the entire field is a major pre-occupation of those involved with precision techniques. There are indeed several ways to assess and depict soil fertility (nutrients) variations. Sensing soil minerals could provide us with rough idea about nutrient status of soil. Simultaneously, measuring soil organic matter (SOM) may be useful, especially to know soil N and C status. We should also note that plant nutrient status is indicative of nutrient availability and general fertility. Obviously, variations in plant nutrient status also suggests about shortages of nutrients, if any. This aspect has been applied rather commonly with major nutrients like N, P, and K. Sometimes Mg and Ca dearth is also sensed using plant analysis. Most importantly, sensor technology that is required during precision farming is available for many nutrients. A sensor-based on chemical analysis and contact is cumbersome to work out during precision farming. Generally, two principles are used to sense soil nutrients. They are diffusion-controlled measurements involving ion-selective electrodes and solid-state electrochemical sensors. There are sensors based on chemical reaction that lead to changes in color or refractive index. It again involves contact with soil sample and information may not be available to the variable rate applicators at a fast pace. In more advanced sensors, soil sampling, and monitoring of soil NO_3 is achieved using ion-selective field effect transistors. During recent years, "on-the-go" sensing of soil nutrients and other properties has been standardized. Automated soil sampling and mapping facilities are available on farm equipment. They provide information on soil N, K, and pH (Hellebrand and Umeda, 2004). Such sophisticated farm vehicles have accelerated evaluation of soil fertility variation and variable rate nutrient inputs into fields.

Soil N status can also be sensed using plant analysis data. Plant leaf color and chlorophyll content are related to nutrient availability in soil. Spectral evaluations of crops are usually confined to N because it is quantitatively most needed nutrient. Its deficiencies occur most frequently. Actually, chlorophyll density in leaves affects inflection point in the red/infra red region of spectra. Spectral signatures of different plant species could also be utilized to arrive at plant N status and later decipher soil N. Another approach is to measure chlorophyll florescence (Bredemeier and Schmidhalter, 2003). Soil N availability may also be estimated indirectly by measuring plant protein content. Spectral analysis of leaf proteins could be made routinely without contact and extrapolated to soil N status. Vegetative index and reflectance measurements also provide data about soil N variability.

The SOM is generally estimated using combustion techniques, but these are not amenable for use on variable rate applicators. Incidentally, soil color, and NIR reflectance are properties influenced by soil organic matter (Sudduth et al., 1990). Hence, it is useful to measure them and decipher SOM status. It is important to understand SOM distribution up to a depth of 30 cm in soil. Therefore, "on-the-go" soil sampling techniques should be appropriately geared.

High resolution spectral analysis using remote sensors or handheld sensors have been used to asses a variety of soil and crop characteristics, although all of them may not be directly relevant to soil fertility and crop response. For example, incidence of disease on crops has been evaluated based on images developed using infrared thermography. Specific fungal diseases like powdery mildew and its intensity could be assessed using infrared thermography.

2.2 SOIL SAMPLING FOR PRECISION FARMING

Soil sampling is an important step during active sensing or chemical analysis and accrual of data. Often, usefulness and accuracy of precision farming techniques is dependent on soil sampling approaches adopted. Soil sampling has been accomplished in a variety of ways based on technology available, purpose and economic constraints. Soil sampling and its analysis to gather relevant data has been a routine exercise for farmers situated world over. However, it is a basic necessity for farmers who intend to adopt precision farming. According to Ferguson and Hergert (2009), historically, soil sampling has been resorted to assess soil nutrient status and moisture content at different depths in the soil profile. Major objective of soil sampling almost remains same for those adopting precision farming. Under precision farming, instead of a large sized field or an area, sampling is done at closer spacing to ascertain within field variations in soil fertility. As stated earlier, soil sampling procedures and analysis are often dependent on soil type, topography, cropping history, crop to be sown, fertilizer and organic manure treatments prescribed, irrigation management schedules, yield goals and profitability. The cost of tedious procedures should be matched by returns from crop yield. An adequate number of samples should be collected at as many grid points and depths possible.

According to Ferguson and Hergert (2009), farmers could consider adopting grid sampling based on following information:

(a) Fields where previous management has altered nutrient status and distribution appreciably. Factors of concern and that need attention are heavy manure inputs, high and uneven crop productivity, erratic crop residue recycling procedures, uneven seepage or percolation of irrigation water and dissolved nutrients, uneven depletion of soil N and other nutrients due to loss to atmosphere and uneven grain harvests.

(b) Merger of fields with different cropping history may result in large scale variations in soil nutrient and moisture distribution pattern.

(c) Most importantly, grid sampling is needed if farmers intend to obtain accurate and precise information on soil nutrient distribution across a field and at various depths in the soil profile. We should note that, currently for major and secondary nutrients, the trend is to prescribe fertilizer after considering nutrients status both at surface and sub-surface layers of soil. Also, we may note that deep-rooted crops like cotton or sugar cane and fruit crops extract nutrients from deeper layers of soil. Management zone sampling could be adopted if information like yield maps, remote sensed images, aerial photography of soils, spatial distribution of soil nutrients and moisture are already available. Management zone sampling suffices if clearly demarcated zones that have consistently supported different productivity levels are available. Management zone sampling can be pursued in situations where fields have not been used consistently for farming crops or livestock and data on natural variations of fertility and moisture is not available.

2.2.1 Grid Sampling

Soil sampling is a necessity in order to prepare representative map that depicts variations in soil fertility. Usually, at least 15–20 sampling sites are identified in a field. Areas suspected with marked variation in soil properties like texture, organic fraction, and high or low fertility status should be sampled. Problem spots like salinity, marshy areas or eroded zones could be avoided. However, if characteristics like salinity or low pH are marked and well spread out, then soil samples have to be drawn specifically from these zones, so that appropriate amendments could be applied.

Basically, there are at least four different ways to sample fields for soil fertility. They are:

(a) Bench mark sampling involves picking soil samples from uniquely different areas within a field (5.0 ha). Unique areas could be identified based on soil type, topography, and cropping systems.
(b) Topographic sampling involves selecting sites for soil sampling based on topography.
(c) Random sampling involves collection of soil samples in a random pattern across a field. Random sampling suits fields that are large and around 30 ha.
(d) Grid sampling is a prescribed method for precision farming. The field is actually sampled in a grid pattern covering points at 0.2 ha in a field of 2–2.5 ha in size. This allows accurate estimation of soil fertility variation in the entire field without bias to any region within the field (see Figure 1). Grid sampling helps us in preparing detailed maps depicting variations for each soil nutrient. Grid sampling is a common practice in many of the cropping zones. New areas brought into cropping and those in continuous cultivation are often sampled using grids. Grid locations, density, and analysis of samples drawn are actually dependent on purpose. Grid samples could be located in a field by taking advantage of previous yield map. It is said that sampling soils at high densities removes inaccuracies that would otherwise occur in a feebly sampled field. During early stages of introduction of precision farming, say in late 1990s, grid samples were drawn from squares representing 4.0 acres, as facilities increased it got reduced to a sample per 2 acres then to one sample each acre. Sampling density does affect the accuracy with which nutrient distribution can be assessed. For example, Ferguson and Hergert (2009) have clearly shown that pattern of soil P distribution deciphered differ enormously, if density or closeness of grid sampling is enhanced by ten folds more than original. Sparse or coarser sampling provided a different picture about soil P distribution to farmers. It is imperative that missed areas in a coarsely sampled field may actually have higher concentrations of P. Influence of density of grid sampling holds true with most, if not all situations for all mineral nutrient distribution and soil characteristics like pH, EC and so on.

Ferguson and Hergert (2009) state that a well done grid sampling and maps prepared using dense sampling can serve farmers as valuable information on soil fertility for many years. Accurately prepared maps with grid locations shown are highly useful while feeding data to variable rate applicators. Optimum sampling density suggested is at least 5–8 samples done per acre, a composite prepared and analyzed to derive data on various mineral contents, pH, EC, CEC, organic matter content, and so on. Fields with relatively more uniform fertility due to consistent use may be sampled sparsely at 2–2.5 acres. Sampling pattern is a crucial exercise. It is often decided based on prior data or coarse knowledge about soil fertility trends in different regions of a farm/field.

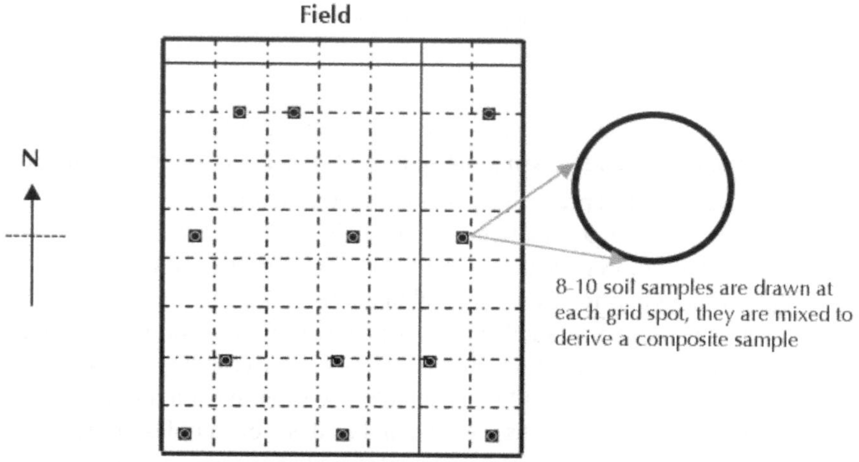

FIGURE 1 Diagram depicts a hypothetical grid markings and soil sampling spots in a Field. *Note*: At each spot in a grid 8–12 soil samples are pooled. A composite sample is then analyzed for soil nutrient status and other characters. Soil fertility maps are drawn and used in the decision support system, so that variable rate nutrient application is accomplished.

Let us consider an example depicting the spatial variability of soil nutrients based on which the intensity of grid sampling could be decided. In Belgium, Gleyic Podzols supporting grass lands and arable cereal cropping zones exhibit a certain degree of spatial variability with regard to soil fertility. The variogram analysis for soil fertility parameters has shown that SOC, pH, Ca show high variation between locations 110–170 m apart. Whereas, for nutrients like K and P variations are smaller. The variation in soil nutrient content was also dependent on cropping systems. For example, variation for K was much greater in grasslands than in cropped fields. Geypens et al. (1999) suggest that, based on prior knowledge about extent of variation for individual or group of essential elements, we can suitably modify grid sampling intensity and other procedures.

We should note that in order to obtain accurate picture of variability of various nutrients, detailed analysis of grid samples is necessary. Distance between samples, depth of soil sampled and number of cores obtained at each spot has a say in the accuracy. This aspect has direct impact on how variable applicators respond to soil data. Standardizations prior to adopting precision techniques are useful. For example, a study reported by Franzen and Swenson (1995), suggest that grid sampling at 110 ft apart with 6–8 core samples composited at each grid point suffices to obtain a good correlation between nutrient content and availability to crops (Figure 1). Surface samples provided good correlation between soil nitrate values and crop response to variable inputs. Smaller grid size and closer cores samples added to accuracy. Similarly, in the paddy production zones of Malaysia, it seems stratified sampling of soil provided better picture about soil fertility variations. The errors were lessened to a great extent, if stratified sampling was adopted. The reduction in standard error of deviation was dependent on number of soil layers sampled and intensity or closeness of sampling spots. Stratified sampling enhanced the accuracy of depiction of soil N and K distribution (Jahanshiri, 2006).

We should note that in addition to precautions regarding grid size and intensity of sampling, nutrient distribution on the surface and subsurface layers of soil needs due attention. In other words, samples drawn from different depths that get pooled later are important. For example, Kasowski and Genereux (1994) found that, if soils were sampled from deeper layers, it affected the fertilizer-N requirement. Fertilizer-N requirement reduced if samples were drawn from 4 ft as well, instead of just 2 ft depth. Further, N lost *via* emissions got reduced, if subsurface N was also considered while deciding on fertilizer schedules for VRT. Revenue increases due to subsurface sampling for N ranged between 63 and 350 US \$ ha^{-1}.

Soil fertility maps prepared using densely sampled grid systems are effective as long term guides during precision farming. Yet, each season soils are refurbished with variable rates of nutrients depending on crop and yield goals. This leads to alterations in soil fertility pattern that was originally prepared using densely spaced grids. Ferguson and Hergert (2009) suggest that soil N pattern and fertilizer-N requirements may alter rather quickly. Whereas, soil maps depicting lime requirements are said to hold good for 8–10 years at a stretch. However, soil maps for immobile elements like P and S may not alter much even after a single or couple of variable rate application events. Soil nutrient buffering and efficiency with which crops remove nutrients is also important. Sampling depth depends on crop species, its rooting pattern, nutrient depletion pattern, physico-chemical traits of soil relevant to localization and availability of mineral nutrients. Fertilizer placement method also influences depth of soil samples drawn in a grid. It is often better to collect samples within a grid region. According to Ferguson and Hergert (2009), grid sampling is not a preferred option for estimating soil nitrate (N) distribution pattern in a field that would support cereal production. Annual fluctuation in soil N is distinct and high enough to make previous reading obsolete. Seasonal or annual grid sampling, chemical analysis and preparation of maps or data to feed the variable rate applicators is cost prohibitive in most situations. Instead, it is customary to sample soils for residual soil N. There are of course other rapid and cost effective methods based on soil sensors, plant analysis (or chlorophyll meter readings) and remote sensing.

Mallarino (1998) explored sampling procedures that suit estimation of soil P, K, and pH variations. The aim was to study the dynamics of P and K in relation to maize crop response. Soil analysis has shown that in Iowa soils is generally optimum or optimum (rich) in P and K. However, spatial variability of P and K in the soil is highly complex. It does cause variability in crop growth. Attempts to find a common soil sampling procedure that suits large areas within the "Corn Belt of USA" seems difficult. Results show that, sometimes, intensive grid sampling provides better judgment about variations of soil P and K. However, sampling by soil type has been commonly adopted in this region. It seems sampling large areas of soil, 3–4 acres in size, does not seem to provide accurate information on soil P and K variations. Sampling large areas seems to overestimate nutrient levels (Mallarino, 1998; Figure 1). On the other hand, reducing sampling distance or cell size increases cost of sampling and analysis. Therefore, it reduces economic viability of precision farming. Hence, it was suggested that obtaining digitized soil maps with information about fertility variations and aerial photographs of bare soil and canopy might be useful. We should note that in soils already optimum or optimum for soil P and K, response to adoption of soil maps and variable rate inputs might be feeble. However, it may have excellent impact on nutrient dynamics in a given field since it reduces on fertilizer inputs. Undue accumulation of soil P and K is avoided due to variable rate supply of these elements. Ideally, soil sampling procedures should be tailored based on exact purpose. Soil sampling procedures for P and K are essentially dependent on factors like soil type, nutrient(s) deficient, extent of variation expected and end use, say to prepare a soil fertility map or to feed data to variable applicator.

Grid sampling, mapping procedure, and deciphering spatial patterns of soil fertility factors are among the most important factors affecting adoption of precision farming and its profitability. Of course, it has direct impact on variable rate inputs; consequently the nutrient dynamics per square gets affected. As stated earlier, intensive grid sampling and soil nutrient analysis could be cost prohibitive. Lakes et al. (2007) suggest that soil sampling strategies could be modified and simplified by using auxiliary data from satellite or aerial imagery. This method of combining a couple of procedures helps in reducing costs on deciphering variations with regard to soil organic matter, soil P and K in fields meant to grow wheat in the next season.

2.2.2 Management Zone Soil Sampling

Management zone sampling is amenable when prior data or yield maps, aerial photographs or knowledge about fertility pattern or disease pressure, or moisture distribution is available. It is generally easy to overlay yield maps or soil fertility maps with aerial surveys and then mark management zones in a field (Ferguson and Hergert, 2009). Sometimes, soil factors like compaction, topsoil depth, subsurface traits, texture, or pH could be criterion to mark management zones. Number of management zones to be created depends on soil fertility patterns known to occur in the field. However, it is advisable to restrict excessive subdivision. Generally, 5–6 management zones should be adequate. We may note that a management zone need not follow any gradient nor does it need to be continuous. For example, within a field, management zone marked based on a particular soil fertility status may occur discontinuously. It may repeat a couple of times per ha (see Figure 2).

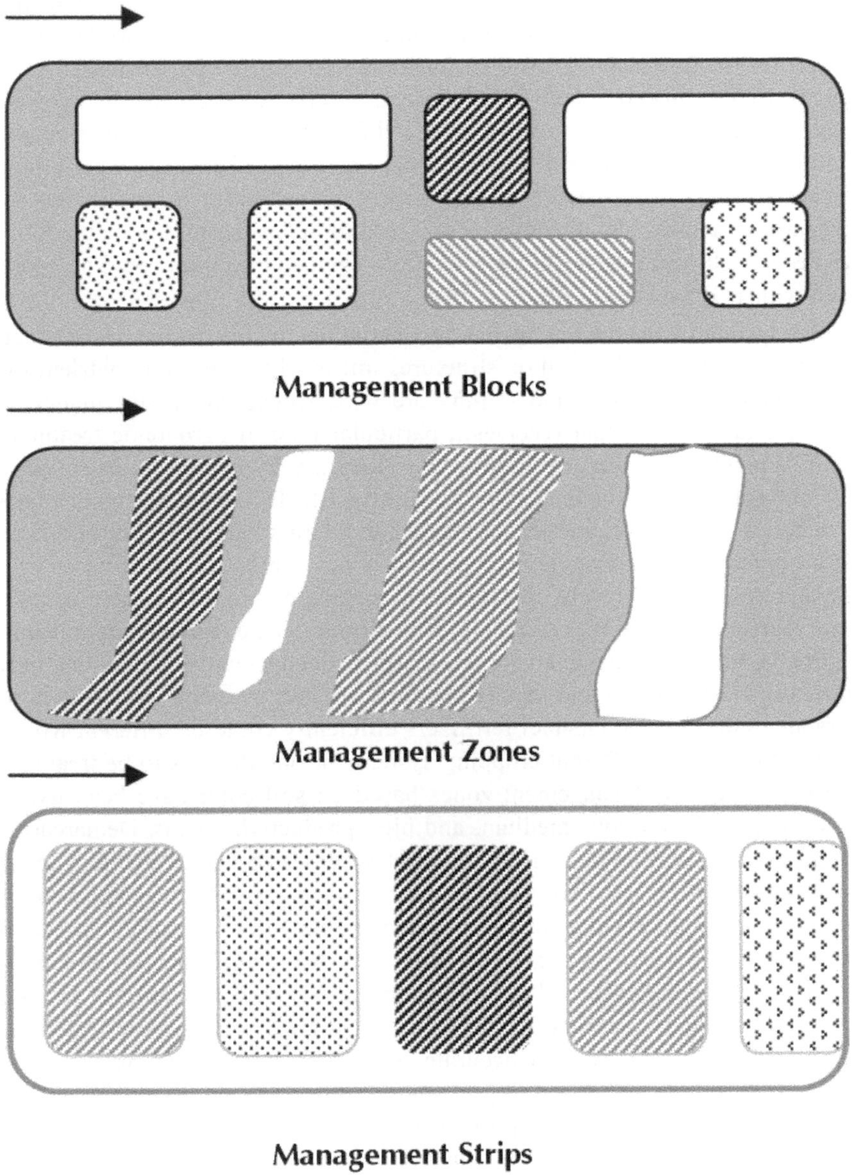

FIGURE 2 Sampling using management blocks, or zones or strips.

Note: Soil samples could be drawn from each block or zone depending on facilities. Soil fertility map for each block or zone or strip is prepared. Results are then supplied to Decision support system or utilized directly under "on-the-go" system. Blocks or zones of a particular level of soil fertility or any other trait may repeat as many times in a field or farm. Arrow indicates the direction of the movement of tractors and variable rate technology fitted vehicles. Management blocks/zones/strips are shown by different color shades and patterns.

Sampling soils at each management zone is necessary. Usually, local recommendations are followed. About 8–10 core samples are collected at surface layers (0–10 cm) and subsurface layers (30–60 cm depth) (Figure 1 and 2). Residual NO_3-N is judged by collecting core samples at 3 ft depth. Soil samples are usually pooled systematically and a composite sample is derived from it. About 6–8 samples get drawn but only composites are chemically analyzed. The chemical analysis is performed on both surface and subsurface soil samples to ascertain nutrient status and fertility levels as accurately as possible. Accuracy of soil fertility maps depends on number of soil samples and stringent chemical analysis. Data from soil samples could be georeferenced with a GPS receiver.

Management zones are needed when variation in soil characteristics that affect crop production like texture, structure, soil fertility especially nutrient distribution, moisture, pH, EC, and salinity are wide spread. Generally, management zone in a farm is one that receives a particular type of agronomic treatment or specified input. For example, sandy or clayey area, a dry land zone, a strip or patch with low soil fertility status, acidity or salinity affected zone and so on. Management zones are highly practical in diversified farms supporting different crops and cropping systems. Simple management zones can be identified based on soil type and productivity levels. Management zones are currently created using data about a series of soil traits, crops, productivity targets, and economic advantages (Figure 3). Actually, we demarcate and classify management zones so that desired inputs can be channeled and results obtained without overlap. Obviously, a management zone meant to channel fertilizers efficiently could be different from that supposed to support different cropping systems or one that has to be treated with lime to correct pH. Management zones based on soil color have been useful in classifying fields into low, medium, and high productivity areas. Demarcation of zones based on soil color and organic carbon has also been beneficial to farmers (Moshia et al., 2010). We should note that management zones are sometimes transitory and need regular monitoring and modifications to match the changes that occur in soil as we cultivate year after year. Although management zones are most practicable and preferred as a procedure under precision farming, we should appreciate that active sensors that assess soil and crop characteristics periodically and identify variations are more accurate. Sensors may lead us to greater accuracy and benefits in terms of uniformity in soil fertility and crop productivity. Moulin et al. (2003) have stated that adoption of variable rate technology was difficult and costly because fields occurring on hummocky terrain of Manitoba in Canada were not classified into simple management zones. Grid sampling was costly and could make the process uneconomical. Hence, they first resorted to methodical cropping and identification of factors that caused soil fertility variations. Then, marked the management zones based on topography, soil nutrient status and yield maps.

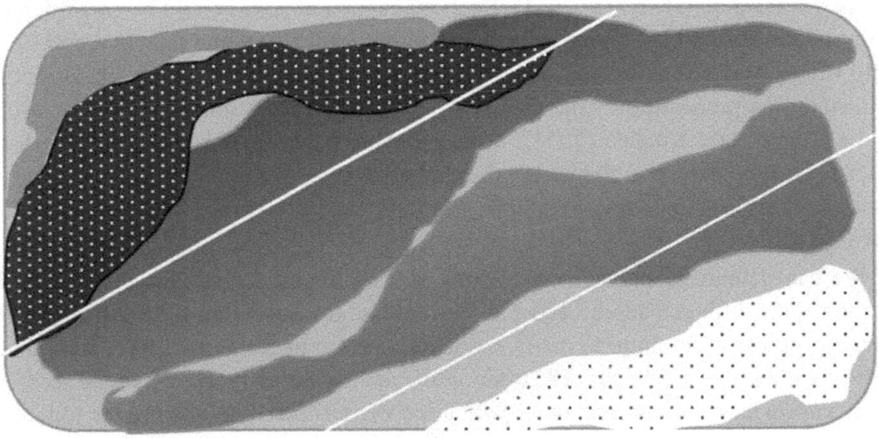

FIGURE 3 Demarcation of management zones based on soil textural variations.
Note: Dark black zones represent predominance of clay content, grey shows higher percentage of silt and white area shows high sand content. Clearly, the field could be classified into three management zones as shown in the figure by white straight lines. Such a demarcation allows farmers to manage fertilizer prescriptions and variable rate applicators more efficiently. For example, management zone with predominance of sand could be provided with greater amounts of organic matter since its paucity is easily felt in sandy soils.

There are several methods that allow us to study variations and demarcate "management zones". For example, in the Australian wheat zone, mature crop is analyzed using digital cameras and data cards are used to set up management zone (GRDC, 2010). In-season variations in crop growth and biomass formation could also be used to decide management zones. Often detailed information about soil and land differences are utilized to set up management zones. Aerial photographs of soil provide good data to demarcate management zones. Actually, soil type or color could be good indictors for boundaries of a management zone. Electromagnetic imaging and maps of EC measurements are also used to decide boundaries of different management zones. Sometimes elevation and topographic data are used as boundaries. Gamma radiometric surveys that show up differences in soil texture can also be used to decide on management zones (Figure 3). Data from soil physical and chemical analysis are among the best to demarcate management zones, especially if the intension is to supply fertilizers at varying rates. Crop variety or disease pressure could also be used to mark zones.

Management zones could be developed for specific applications, say to target N supply. (Franzen and Kitchen, 2011). Such management zones could be constructed using the usual data like, topography, aerial photographs, satellite imagery, soil EC, yield maps, and intensive soil survey data. Actually, several layers of data are utilized during preparation of management zones meant for N management. According to Franzen and Kitchen (2011), often more than one layer of information that gets used, coincide and authenticate the demarcations. Most importantly, management zones are specific, easy to handle, accurate and allow precision applicators to operate swiftly.

We can demarcate management zones to specifically supply variable rates of fertilizer-P at different depths and reduce its loss. Encountering fields with soil P related problem is common to many cropping zones. Snyder et al. (2011) state that in the Northeast of USA, farmers often come across fields those are prone to soil P loss *via* surface flow or erosion. Application of fertilizer-P to such fields reduces efficiency of inputs rather drastically. Hence, it was suggested that zones prone to surface runoff and erosion could specifically marked into management zones and fertilizer-P supply could calculated based on extent of loss encountered and suitable measures could be adopted. Deep placement of fertilizer-P specifically in the runoff prone zone using suitably guided variable applicators improves fertilizer-P efficiency.

Let us consider an example, where in, management zones were demarcated based on a soil physico-chemical property such as EC. Actually, management zones marked based on soil apparent EC could be a reliable indicator of potential yield obtainable in the area. According to Aimrun et al. (2011), soils from fields that support paddy could be classified and management zones marked based on apparent EC. The management zones that could be identified into Zone-1 with shallow EC had significantly low EC, Ca K, and Fe, but relatively higher fine sand. Zone-2 had deep EC and lower level of Mg, Na and total CEC. It has been suggested that field scale delimitation based on EC could be useful to farmers in channeling fertilizers and other inputs using variable rate applicators.

2.2.3 Strips and Field Plots

Management strips and fields may be easily created using criteria such a cropping pattern, soil characteristics like texture, color or fertility (Plate 1).

Such demarcations are quite frequently adopted when crop species sown are different and are intended to match variations in geographic traits, soil characteristics and yield goals. Management strip allows farmers to stagger and/or separate out various farming procedures. It allows farmers to sample soils, estimate nutrient and moisture status more accurately. Adoption of variable rate seeding, nutrient and irrigation is easier if large farms are demarcated into smaller strips or fields.

PLATE 1 Top: Management strips created based on soil type and cropping pattern.
Bottom: Strips created based on black soil.

Note: Strips of maize or soybean are managed relatively easily during soil sampling, fertilizer supply, pesticide application and harvesting.

Source: Mr David Nelson, Nelson Farms Inc, Fort Dodge, Iowa, USA.

2.3 SENSORS AND ESTIMATION OF VARIABILITY WITHIN CROP FIELDS

2.3.1 Remote Sensing: Active and Passive Approaches

During recent years, techniques such as active soil sensing, satellite photography and multispectral imaging that are generally associated with futuristic space and satellite-based communication are being increasingly introduced into farm management. Now, they are actually considered valuable and futuristic aspects of agriculture (Sonka et al., 1998). Remote sensing is an innovative approach that allows us to assess soil, crop growth and to judge precise requirements of fertilizer, irrigation, and other agronomic procedures like pesticide application. Remote sensing is actually a process of gathering information of an object (e.g. soil or crop) without being in close contact (Khosla, 2008). Both passive and active remote sensors are being used during precision farming. Passive remote sensors are usually mounted on satellites, airplanes or balloons. The passive remote sensing satellites cannot acquire information on soils or crops during night. Weather parameters like clouds may also affect accuracy. Studying the imagery and interpretation is yet another problem attached with passive remote sensing. Passive remote sensing usually utilizes light wavelength in the visible band for assessing crops. The NIR light is also used to detect soil water status, crop disease or nutrient status.

"Active Remote Sensing" or Sensor-based crop and soil assessment is popular with farmers adopting precision farming. It is a latest technique that overcomes certain limitations attached with passive systems mounted on satellites. Firstly, active sensor has its own light source. Hence, it is called "active". It allows data accrual both during day and night. It measures crop/soil reflectance both in visible and non-visible range of electromagnetic spectrum. Active sensors are hand-held, inexpensive gadgets easily procurable even by resource poor farmers. Most importantly, the spatial resolution of hand-held sensors (active) is quite high, easy to decode, interpret, and quantify. Since the sensors are held closer to crop canopy or soil surface, details on imagery are excellent and most accurate. There is indeed a wide range of crop canopy sensors available in market to suit the purpose and monetary resources of the farmers (Khosla, 2008). Several of the models are easily mountable on tractors and other farm vehicles.

2.3.2 Sensors Utilized to Evaluate Variations in Soil Fertility

Firstly, precision farming or site-specific management aims at accurate management of soil fertility, moisture and other inputs based on grids. Therefore, detailed knowledge of soil and crop characteristics that occur within each grid is required. Generally, it is too tedious to sample and chemically analyze soil samples collected from several hundreds of locations in a field. At times, soil analysis is cost prohibitive and time consuming. Results from soil tests may not be available on hand before seeding. Therefore, demarcating fields into subplots or grids and scanning the soil or crop with appropriate sensors is pertinent. Sensors could be accurate, rapid and helpful in feeding the data into variable nutrient applicators in time, as and when the crop is seeded. A wide range of sensors is used during precision farming. Sensors accomplish several different operations in the field (Upadhyaya and Teixeira, 2010). Sensors are highly specific and built based on exact task for which they are meant. According to a recent review by Kim et al. (2009), among the variety of sensors that have been developed and used, it is generally believed that "on-the-go" vehicle-based sensing systems would be preferred over others.

Following is a list of parameters and functions for which sensors have been used:

Soil Sensors: Soil texture, structure, physical conditions, moisture distribution, nutrients in profile, EC, and so on.

Crop Sensors: Plant density, nutrient status, moisture stress, disease incidence

Yield Sensors: Monitoring grain or fruit yield, forage yield, and so on

Variable rate Technology Systems: Fertilizer flow, water flow, and weed density.

Source: Upadhyaya and Teixeira, 2010

Following is yet another way of categorizing sensors used to assess soil and plants:

Based on Principles used in Sensor Technology: Electrical and Electromagnetic, Optical and Radiometric, Mechanical, Acoustic and Pneumatic, and electrochemical.

Based on targeted Properties estimated by Sensors:

Plant Sensors: Biomass, Nutrient status (deficiency), Weed population, Water stress

Soil Sensors: Texture, Organic matter, Moisture, Nutrients, soil pH, and Compaction.

Source: Dobermann et al. 2004

Sensors are built based on different physico-chemical phenomena. The most common technology adopted in sensors is based on electromagnetic induction, EC, ion selective field transistors, optoelectronic sensors, ultrasonic displacement, and visual systems. Quite often, a combination of techniques is employed in order to make the sensor most suitable to a variety of situations and to gain in accuracy. For example, in a combine harvester, usually contact sensors, ultrasonic sensors, pressure, linear displacement, and opto-electronic sensors are used to assess grain quantity harvested. Dobermann et al. (2004) have pointed out that commercialization of "on-the-go" soil and crop sensors began earnestly during early part of 2000 A.D. At the same time, electrical and electromagnetic sensors for soil mapping became most sought after items. Several companies like Veris Technologies, Geonics Ltd, Geocarta Inc and Crop Technology Inc., began producing soil sensors. These gadgets allowed estimation of pH, EC and nutrients in the soil. Crop sensors that helped in variable rate N inputs were also developed (e.g. Hydro-N sensor). Sensors that differentiated and estimated weed intensity were also developed during this same period.

Sensors to Estimate Soil Properties

According to Chung et al. (2008), soil properties such as texture, moisture, pH, EC, nutrients like Ca, Mg, Na, N, P, and K could be assessed using closely placed sensors that operate at visible and near-infrared wavelength (Plate 2 and 3). They can gather data rapidly from soil samples derived using cores at different depth and locations. Further, such data can be easily loaded on to electronically operated variable rate applicators. Such site-specific methods impart uniformity to soil properties and allow us to channel inputs based on yield goals. Generally, it is worthwhile for the farmers to obtain a soil fertility map or soil map depicting various physico-chemical factors (texture, pH, CEC), major nutrients and soil moisture. Knowledge about critical values for the soil parameters and ability to correct it "on-the-go" or through the use of previously prepared maps seems important (Johnston et al., 1998).

PLATE 2 A Soil Sensor that measures and maps pH "on-the-go".

Note: Prior knowledge about soil pH and its variations within fields is important. Management zones and cropping pattern could be marked based on soil pH. Soil pH affects rooting, nutrient availability and recovery by the crop.

Source: Mr. Eric Lund, Veris Technologies Inc http://www/veristech.com/products/soilpH.aspx (August 1, 2011)

Electrical Conductivity

Each of the soil characteristics like nature and quantity of clay, salinity, cation exchange capacity (CEC), soil moisture content, depth of clay pan and abundance of specific ions has its proportionate effect on EC of soil and consequent crop growth. We should note that expression or quantity of each of the soil factors varies enormously in soil and in a given zone. Therefore, variation in EC needs to be mapped, before adopting it in precision farming. Generally, there are two methods that help in estimating EC. They are instrumentation based on electromagnetic induction and those that measure EC directly (Upadhyaya and Teixeira, 2010).

Doerge et al. (2011) have put forth following arguments regarding usefulness of mapping EC to assess soil traits, fertility and crop productivity.

They are:

(a) Soil EC mapping is simple to read and infer. It is inexpensive. It can be used quickly and accurately to characterize soil differences.
(b) Measurements of soil EC correlates to soil properties that influence soil texture, CEC, organic matter, salinity and sub-soil characteristics.
(c) Most importantly, soil EC is related to specific soil properties that affect crop yield, such as soil depth, pH, salt concentrations and water holding capacity.
(d) Interestingly, soil EC maps coincided with crop yield patterns. It can even be verified quickly visually using EC map of field. The soil EC data also correlates with those from remote sensing.
(e) Soil EC maps may also help us in indicating variations in natural resources, general vegetation potential of an area. It may be used in preparing schedules for remediation of soil salinity, drainage, and so on.

PLATE 3 A soil electrical conductivity mapper.

Note: Soil EC Mapper detects changes in soil EC and maps it "on-the-go". Soil EC is supposedly an indirect measure of soil productivity.

Source: Mr. Eric Lund, Veris Technologies Inc. http://www/veristech.com/products/soilec.aspx (August 1st, 2011)

Sensors for Soil Nitrogen

During conventional cultivation of cereals or other crops, applying fertilizer, rather uniformly, based on state agricultural agency recommendations and yield goals, usually satisfies N requirement. This is done despite clear knowledge that soil N is a highly variable factor. Further, we ought to realize that rates of N transformation processes like mineralization, immobilization and nitrification are all variable. They are affected by a wide range of soil and environmental factors leading to variable availability of N to crop roots. Loss of soil N through natural process like percolation, erosion, seepage and emissions are also highly variable. Therefore, soil N measured at seeding may not be a relevant measure. Calibrations that account for various losses and transformation need to be in place within various computer models used to recommend fertilizer-N. Fertilizer-N is usually applied in splits during intensive cultivation of major cereals and cash crops. This improves its use efficiency. According to Upadhyaya and Teixeira (2010), sensors and instrumentation that aid efficient and variable rates of fertilizer-N supply into a field is one of the most important aspects of precision farming approaches. It has been suggested that near-infrared reflectance could be used to determine soil mineral-N content. A multiwavelength NIR absorbance spectroscopy could help in mapping soil NO_3-N variation. Real time NO_3-N sensors developed using Ion selective electrodes are available. Time required for NO_3-N extraction seems to be a constraint for use in automatic soil N sensors (Upadhyaya and Teixeira, 2010). Hand-held soil N sensors are available (e.g. Spectrum technologies Inc.).

Multiple Sensors to Estimate Soil Physico-chemical Properties

As stated earlier, precision farming involves collection and analysis of relatively larger number of soil samples. Further, tedious chemical analysis takes time and results may not be available in time to run a variable applicator. Regular chemical analysis of large number of soil samples is also cost prohibitive. Sensors that make a rapid estimate of several soil properties, at a time, is of course a preferred solution (Plate 4 and 5). According to La et al. (2008), there are sensors that operate based on optical reflectance. Sensing is done in the visual and infrared wavelength bands. There are also a set of sensors that utilize selective electrodes to estimate soil chemical properties. Optical sensors are effective in assessing soil physical properties but are ineffective in providing useful data on soil mineral nutrient status, say, P or K content of soils. On the other hand, sensors based on Ion Selective electrodes provide accurate estimates of nutrients like P and K, plus other soil chemical traits like pH, lime and so on. Therefore, La et al. (2008) have tried to fuse the two principles into a single instrument (sensor) and make a rapid and cost effective evaluation of variations in soil characteristics. Such multiple sensors could be very useful in providing data for use in marking management zones and in deploying variable rate applicators. Results reported by La et al. (2008) indicate that good estimates (R^2 = 0.83 to 0.93) were obtained from spectral data for items like soil texture, its fractions, organic matter and CEC. Estimates on pH, P, and K were inconsistent using spectral data. However, when Ion-selective systems were fused, information on P and K were accurate and highly correlated (R^2 = 0.93) with regular chemical analysis. Multiple sensors have been tested in many of the European cropping zones. In Spain, researchers believe that current trend is to adopt sensors that allow rapid and continuous collection of data about the soil. This helps farmers in using on-the-go precision techniques. Lemos et al. (2007) report that sensors that evaluate soil pH and Ca content have performed satisfactorily providing excellent correlations with standard manual estimations. Actually, "on-the-go" sensors could hasten certain farm operations relevant to fertilizer supply.

According to Lee et al. (2007), sensors that estimate soil properties and identify spatial variations without need for extensive sampling is a promising alternative. In this regard, optical reflectance sensing in visible and near infrared wavelength bands has received most attention. It was found that Visible-NIR reflectance estimated surface characteristics like clay, Ca, CEC, and Organic-C satisfactorily. They believe that sensors based on visible-NIR band may be useful in assessing soils (Plate 5).

PLATE 4 A Multiple soil sensor to detect and map soil physico-chemical properties.

Note: Such soil sensors possess ability to map multiple properties like EC, Texture and Organic matter. They are based on both optic and electrical properties of soil.

Source: Mr. Eric Lund, Veris Technologies Inc, http://www/veristech.com/products/opticmapper.aspx (August 1st, 2011).

Let us consider an example. Christy et al. (2010) have examined multiple sensors that could be used during "on-the-go" soil analysis and variable rate treatments. The sensors described measure soil reflectance at wavelengths from 9501650 nm. Such infrared sensors have been used to map the soil by analyzing samples at 20 m transects. Validations through field trials in Central Kansas have shown that NIR-based multiple sensors provide excellent estimates of organic matter, organic-N and buffering capacity. Measurements were moderately accurate for soil pH and soil P (Mehlich's-P).

Bogrecki and Lee (2007) compared soil P sensors that were used on ultraviolet, visible and near infrared radiations. Sensors based on soil reflectance from infrared sources were more accurate. Maleki (2010) states that among the sensors, those useful during "on-the-go" variable rate P supply will most sought after. Three basic requirements to be accomplished are:

(a) To gather soil P status while the traction machinery on the move,
(b) Computer model that allows us to decipher soil P content, and
(c) A computer model that develops most appropriate recommendation for P input at a particular spot.

Soil organic carbon (SOC) is an important component that affects soil quality, nutrient transformations, buffering and availability of essential nutrients. The SOC content varies enormously, based on variety of factors related to soil formation, inorganic fertilizer inputs, cropping systems, organic manure supply and recycling of crop residues. Maps depicting variations in SOM can be prepared, although it is slightly tedious. The organic matter can be measured using thermal and chemical techniques. Usually combustion of dry soil sample and measurement of mass change or CO_2 gravimetry is done. ^{13}C carbon isotope based techniques are also common. It allows us to get an estimate of CO_2 concentrations in soil. Yet, the two methods are not amenable to be used on variable rate applicators that have to decide organic matter supply rather quickly in response to SOM changes. However, there are techniques based on CO_2 analysis and near-infrared (NIR) reflectance data could also be used to operate variable rate applicators. We should note that, color and NIR reflectance are influenced by organic matter content. Therefore, optical methods that allow estimation of SOM are preferred (Rossel et al., 2003; Selige et al., 2003). Application of optical methods is difficult when factors interfering with SOM are conspicuous. Sometimes, soil depth, surface features, water content, mineral content may also affect accurate measurement of soil organic carbon.

2.3.3 Crop Reflectance Sensors

Crop reflectance sensors depend on spectroscopic analysis of light reflected from crop. They are used to assess and monitor crop growth, leaf-N status and infestation by insects or plant pathogens. Influence of soil fertility and maladies on crop growth can be assessed and mapped using crop reflectance sensors. Currently, agriculturists utilize data from both, remote sensing and hand-held sensors to develop variable rate maps.

Sudduth et al. (2007; 2010) have made use of a series of crop reflectance sensors (e.g. GreenSeeker, CropCircle and CropSpec) to assess nitrogen needs of maize crop grown in the plains of Missouri, in USA. They suggest that development of appropriate software that converts reflectance values to plant-N status and development of commensurate control plots that allow assessment of sensors plus variable applicator effects on soil N and crop growth effects is crucial.

PLATE 5 An optic multiple soil sensor.

Note: Such Multiple sensors are often based on Visual-Near Infra-Red Spectrophotometry and Electrochemical phenomenon. They measure and map Soil C, N, Mg, K, pH, and Illinois Soil N test values.

Source: Mr. Eric Lund, Veris Technologies Inc. http://www.veristech.com/products/visnir.aspx (August1st, 2011)

In the Corn Belt of USA, precision farming has mainly aimed at improving the efficiency of fertilizer-N and water. Methods that economize on N supply or improve crop response to N are sought. In this regard, Shaver et al. (2010) has reported that it is possible to estimate variations in availability of N to corn crop grown in semi-arid regions of USA, using crop canopy sensors. Reflectance data derived from crop canopy sensors (ground-based) have been useful in diagnosing crop-N requirements at different stages of the crop and at all locations in farms or small regions. This is a specific advantage of ground-based and hand-held crop canopy sensors over wide range traditional remote sensing. Crop canopy sensors are being routinely used to assess temporal and spatial variations in N status of the crop grown in different climates.

Let us consider a few more pertinent examples. In Argentina, active sensors for reflectance measurements of maize and wheat are used to direct fertilizer-N, during vegetative growth stages of the crop. They actually convert sensor data into normalized difference vegetative index (NDVI) and chlorophyll index (CI590) (Solari et al., 2008). Schmidt et al. (2009) suggest that sensor reflectance could be used to assess to N requirements of maize more accurately at 6th or 7th mature leaf stage.

Identification of variability in crop N status within a crop season and fixing split N application are important aspects of cereal crop production in the Great Plains of USA. There are several improvements that have been tried and tested to diagnose crop-N status in order to decide the quantum of fertilizer-N that could be supplied in-season. Nitrogen diagnostics using hyper-spectral line imaging is known to avoid certain problems otherwise associated with non-imaging low spectral resolution systems. According to Jorgensen (2002), hyper spectral line imaging system (e.g. 'VTTVIS- Imspector V7' by Specim Ltd, Finland) allows rapid collection of spectral information from thousands of spots. It overcomes certain problems related to confounding factors like water stress and its effect on crop growth. Such Inspector V7 based data can predict chlorophyll content and leaf-N concentration with greater accuracy.

Hyper spectral radiometric analysis of crop canopy provides a quantitative estimate of plant-N concentration (Stroppiana et al., 2009). Spectral data derived from hand-held spectroradiometer could be effectively used to assess nitrogen/chlorophyll content in a paddy crop. Spectral data derived from a range of bandwidth in blue/green range spaced at 10 nm was utilized to derive estimates of vegetative index and chlorophyll content. A graph that has vegetative index regressed against plant-N concentration could be used to estimate plant N. This information could then be transmitted to a variable-rate N applicator in paddy fields. According to Stroppiana et al. (2009), precision and accuracy derived from such a spectral analysis linked to variable rate N applicators is good enough to correct temporal and spatial variations that occur in fields

Sensor technology has been adopted to quantify N status of cotton leaves. The sensor readings have been calibrated and used to supply fertilizer-N. Nutrient supply based on sensor readings has been in place in the Coastal plains since past 9 years. It is believed that sensors may help us to reduce fertilizer inputs and improve cotton crop yield (Balkcom and Rodekohr, 2010).

Multispectral analysis of canopy has been used to assess plant-N status in the rice growing regions of China. Field trials using NIR has shown that spectral reflectance measurements and plant N status correlate. Further, results indicate that canopy reflectance converted to vegetation index (RVI) and normalized difference vegetation index provide better prediction of rice crop's N status (r^2 = 0.820.94) (Zhang et al. 2006). A similar study by Huang et al. (2003) has shown that rice plant's N status could be deciphered using remote sensing and multispectral images. Multispectral imaging using four different band wavelengths at 555, 660, 680, and 780 could be successfully utilized to detect plant's N status. Calibrations indicated that regression coefficient (r^2) was 0.78. Similar results were obtained when spectral scanning was done using canopy sensors placed 15 m the crop. Most importantly, data obtained using remote sensing and and/or sensors placed close to crop could be supplied directly into variable rate N applicators. Such sensor based supply of fertilizer N is becoming common in different cropping zones of China.

Digital Aerial Imagery to Estimate Nitrogen Requirement of Crop

It is difficult yet possible to standardize and use digital aerial imagery to monitor crop-N status and occurrence of N deficiency. Generally, crop-N status is dynamic and rapid changes in crop-N concentrations are difficult to trace, in order to prescribe in-season variable rate N supply. For quantitative use, late season digital imagery seems more accurate and apt. During 2006 to 2009, Kyveryga et al. (2010a) evaluated corn crop canopy at several locations, about 690 in all in Illinois, USA, using color and near infrared digital aerial imagery. They compared values derived from corn stalk-N status with results from digital imagery. The digital imagery was actually enhanced using normalized reflectance values. According to them, lighter color indicated N deficiency in corn plants. The green band predicted and correctly identified deficiency/sufficiency of corn N status in about 70% of the samples. Hence, they suggested that late season corn could be monitored using enhanced digital imagery techniques and variable-N rates could be supplied based on the data.

Corn Stalk Nitrate Indicates Nitrogen Sufficiency Range

Kyveryga et al. (2009) have highlighted the use of corn stalk N as a guide to plant-N status and fertilizer-N inputs. They have argued that distinguishing and/or estimating of N derived from soil N pool and that supplied *via* fertilizer is difficult. These fractions are prone to interaction based on soil moisture availabilities and its variations. Factors like ambient weather, SOM, mineralization rates and loss from profile may also affect recovery of N by plants. In general, techniques that allow feedback on extent of N accumulated in maize plants are necessary. It then helps us in deciding split applications of fertilizer-N. In this regard, measurement of corn stalk-N is said to be highly useful in guiding fertilizer-N requirements during in-season split application. Previous data regarding N distribution in the field and crop productivity maps are helpful in guiding sampling regions for stalk-N estimations. According to Kyveryga et al. (2009), recent techniques such as remote sensing or aerial imagery can provide valuable feedback information about N status and its variability in the entire field. Arial images drawn at the end of growing season are useful in guiding sampling regions within a field. Aerial imagery also helps in categorizing fields based on nutrient sufficiency index values. Soil management histories available with farmers could also be used to demarcate corn stalk sampling locations.

Chlorophyll Meter (SPAD) to Estimate Leaf-nitrogen status

Precision farming involves accurate estimations of soil and plant nutrient status. However, plant nutrient status is actually influenced by a variety of factors related to soil fertility, plant physiological processes and ambient atmosphere. Knowledge about spatial and temporal variations in plant nutrient status, especially major nutrients like N, P, and K is essential to calculate nutrient requirements of different regions in a field and operate a variable rate fertilizer applicator. Grid sampling and tedious chemical analysis procedures may not be an option, if fields are small. In fact, farmers endowed with only small farms may not be able to resort to repeated plant analysis at different stages to ascertain temporal changes in plant nutrient status and feed the necessary data to the variable rate applicators. For example, farms supporting cereals like sorghum or millet in dry lands or wheat on small farms in Southeast Asia may not opt to repeated plant sampling and analysis.

Nitrogen is a nutrient required by crops in higher quantities. Its supply has direct influence on crop growth and grain yield. However, in a given field, N status and requirements of major cereals like rice, maize, sorghum or wheat varies perceptibly, both spatially and temporally. Crops well nourished with fertilizer-N produce healthy green leaves with optimum levels of chlorophyll. In fact, estimation of chlorophyll is ordinarily done to ascertain leaf-N status and plant health. The chlorophyll meter or soil plant analysis development (SPAD) is a simple diagnostic instrument. It is available in hand-held portable versions that are economically affordable for even resource poor farmers. Chlorophyll meters measure the relative green color or chlorophyll content of leaves that has direct correlation with leaf-N status (Plate 6).

Actually, linear relationship of chlorophyll content with leaf-N status has been the basis for development of this instrument (Peng et al., 1995; Figure 4). In large farms, computer aided rapid calibration helps in converting SPAD-chlorophyll meter readings into leaf-N status. Then, suitable crop growth models calculate the fertilizer-N to be placed in the field at each spot.

In small farms, farmers can prepare a field map depicting SPAD chlorophyll readings or plant-N status and manually apply variable levels of N accordingly. The SPAD readings have been used on a wide range of crops in different cropping zones. For example, it has been effectively used on rice in Malaysia. According to Gholizadeh et al. (2009), since SPAD readings are closely related to leaf-N status, SPAD-chlorophyll meters could be used to monitor plant-N status and calculate N requirements. Fertilizer-N could be channeled accordingly at different growth stages of the crop. This way it enhances the fertilizer-N use efficiency significantly compared with uniform or single rate N application. Currently, hand-held SPAD-Chlorophyll meter or plant-N sensor seems most popular and frequently opted instrument by farmers who own small farms. It helps them to identify soil N fertility variation and fertilizer-N requirements based on yield goals. Use of SPAD chlorophyll meter readings is known to improve fertilizer-N use efficiency of high yielding rice, grown in many Southeast Asian countries (Peng et al., 1996; Figure 4). Incidentally, sensors that estimate grain N and protein content are also available for use during precision farming (Thylen et al., 2002).

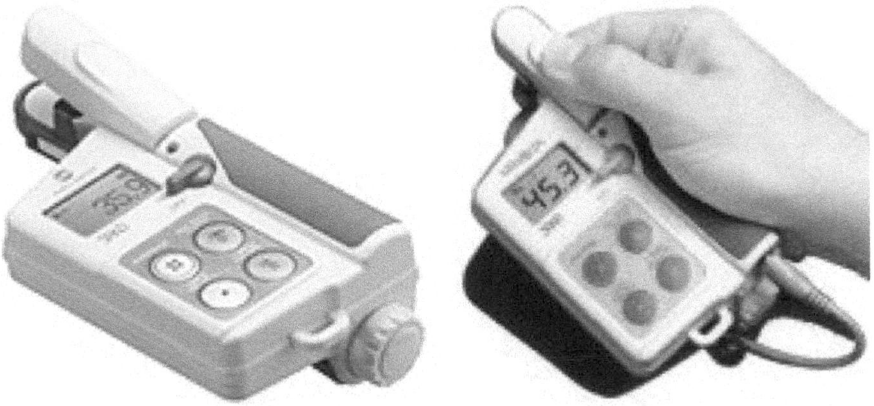

PLATE 6 A Hand-held SPAD chlorophyll meter model 502 Plus.

Note: Such hand-held sensors allow farmers to assess leaf-N status and then decide on fertilizer supply to the crop. It also helps in estimating Plant-N and Protein content. The data accrued could be utilized in preparing maps showing plant-N status or could be used in feeding Variable rate N applicators.

Source: KonicaMinolta Sensing Inc. Osaka, Japan; http://konicaminolta.com/instruments/about/outline.html (July 13th, 2011).

Sensors to Estimate Plant Population

Crop sensors that estimate planting density at various stages of growth are available. They are suitably modified to estimate population of a wide range of crop species. Let us consider a few examples. An optoelectronic sensor was developed and tested by Plattner et al. (1996). The sensor consisted of photoelectric emitter and receiver pair that deciphers plant-to-plant distance within a row. The sensor had to be mounted suitably at vantage point or on a combine to estimate maize seedling population. Maize population was estimated with an accuracy of ± 3.1% at early growth stage and at ± 6.2% at ripening stage. Estimating plant population has definite advantages in forecasting cob/grain and forage yields. It also allows us to study the effects of interplant competition for soil moisture and nutrients. Plant density has direct impact on rooting, nutrient absorption and accumulation rates. Plasticity in crop growth and yield formation that is commonly observed in densely or sparsely planted fields could also be studied using such sensors. Fertilizer-based nutrient supply definitely depends to a great extent on planting density adopted by farmers.

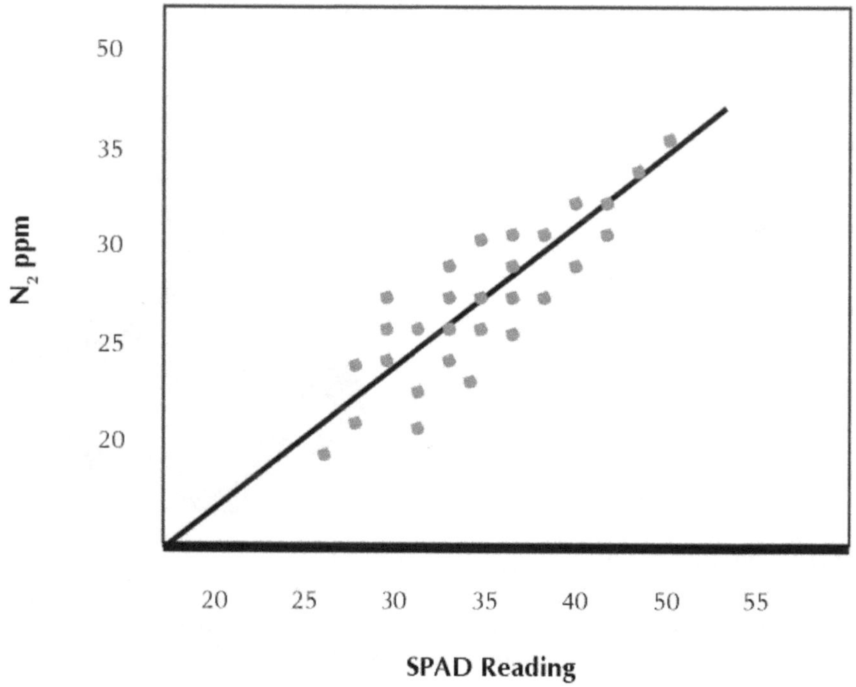

FIGURE 4 A hypothetical graph depicting relationship between Nitrogen content in flag leaf of a cereal plant and SPAD reading from a chlorophyll meter.

Note: Linear regression with high R^2 values of > 0.9 is common. Stage of the crop and leaf position is important. Often flag leaf from seedling aged 30–80 days after planting is selected in case of cereal crops like rice, wheat or maize.

Sensors to Measure Photosynthetic Activity

A optical sensor designed to measure photosynthetic activity *in situ* has been explained by Kebabian et al. (1998). The sensor operates on the principle that as light collected from fluorescing plants is passed through a cell containing oxygen at low pressure, oxygen absorbs the energy and subsequently re-emits photons that can be detected using a photo-multiplier tube. There are several models and variations of sensors that measure photosynthetic activity (Plate 7).

PLATE 7 Hand-held photosynthetic system model CI-340.

Note: Such Hand-held sensors are capable of accruing data about CO_2 levels, luminescence, leaf temperature and stomatal conductance at one stretch rather rapidly. Data could be utilized both, to prepare digital maps and to feed variable rate applicators.

Source: Mr. Michael Larman, CID-Bioscience Inc, Camas, Washington, USA.

2.3.4 Sensors to Monitor Crop Yield (forage/grain)

Yield monitors are used to assess the variations in crop growth and yield. Obviously, prior knowledge about factors that induce yield variations is useful. Factors that most commonly affect crop growth and yield or those which cause variations to crop yield are as follows (see Atherton et al., 1999):

(a) *Soil Factors*: Soil factors that cause yield variations are moisture, aeration, nutrient status, pH, depth of soil, EC, soil microbial load, and composition.

(b) *Topographic Factors*: Relative elevation of the field, physical properties of the soil, slope of the field, contour bunds, locations of drains and so on.

(c) *Climatological Factors*: Precipitation pattern is an important factor. It affects soil moisture distribution. In some of the intensive farming zones like Corn Belt of Iowa, soil moisture distribution is said to explain 69% of crop yield variability. Interactions of soil fertility with moisture is said to be the most important factor that contributes to most of the yield variations. A few other climatological factors like diurnal length, temperature, relative humidity, light interception, dust storms, and soil erosion may also contribute simultaneously to yield variations.

(d) *Cultural Factors*: Aspects like planting density, planting dates; intercropping trends if any, intensity of various routine agronomic practices, fertilizer application schedules, irrigation, and crop genotype may all affect uniformity in crop yield.

The basic purpose of monitoring crop growth, its nutrient status and forage/grain yield formation at different stages is to establish a map that depicts variations in crop productivity. In case, maps depicting variations in soil fertility or nutrient availability are available, we can superimpose them and compare the effects of soil factors with yield maps. Then, devise suitable remedies. Yield maps from previous season could be put to good use while feeding electronic information into variable rate applicators. Yield maps provide us with a general idea regarding soil fertility needs and in fixing yield goals. Electronically controlled yield monitors can be used on crop that is harvested using combines. However, yield can also be monitored and mapped when farmers harvest forage/grain using simple mechanical devise or by hand manually. Factors like crop species, specific product and ease with which it can be quantified may affect yield monitoring process. For example, grains/forage from cereals or legume field crops can be easily monitored. Fruits that are handpicked may have to be appropriately quantified. Large scale wheat or maize or rice production zones could be easily monitored for yield by using combine harvesters fitted with electronically controlled devices that quantify grain yield accurately and prepare maps instantaneously in the field (see Campbell, 1998). Depending on nature of crop yield in question, instantaneous yield monitoring basically involves measuring flow of grain/forage using sensors, grain moisture sensors, ground speed sensors and a computer display. When combine harvesters are used, yield in a management zone or strip or grid could be assessed adopting following equation:

$$\text{Grain Yield} = 14.85 \times FR/GS \times W$$

where,

Yield = Crop Yield (t ac^{-1}); FR = Flow rate (lbs sec^{-1}); GS = Ground speed (miles hr^{-1}); W = Harvester width

Note: Numerical constant is specific and depends on units employed.

Source: Hall et al. (1998)

Yield Monitors

The potential benefit and risk from yield monitor has been summarized Doerge (1999). Basically, yield monitoring and mapping provides farmers with a whole farm perspective about usage of different agronomic procedures and crop management decisions. Whereas, a variable rate technique allows farmers only a good idea about the effects of inputs within a field. Yield monitors were originally developed to assess productivity of large expanses of cereals and oilseeds that occur in Great Plains of North America. However, yield monitors are used routinely on other crops like sugarbeet, tomatoes, peanuts and grapes (Lowenberg-DeBoer, 2003a). Horticultural orchards like apple and pears also use yield monitors. According to Lowenberg-DeBoer (2003a), use of yield monitors is substantially high in large maize farms. Surveys indicate that 60% of Corn Belt farms utilize yield monitors. About 8% of potato farms and 2% of sugarbeet area are assessed using yield monitors.

Yield monitors are used in the Pampas of Argentina, mainly to assess large expanse of wheat, maize or soybean. Long harvest season, large combines fitted with yield monitoring technology and custom harvesting allow the farmers to cover 5,000 ha in a season (Bongiovanni and Lowenberg-DeBoer, 2005). Major constraints to use of yield monitors are high initial investment, lack of management induced variability of soil, low use of fertilizers and risk due to grain price fluctuations. Lowenberg-DeBoer (2003a) states that yield monitors become economically viable, if large farms of 1,000 plus ha can obtain 0.5 q grain ha^{-1} more than conventional systems. The extra grain or reduction in fertilizers pays for purchase of yield monitors.

Casady et al. (2010) have recently reviewed information about yield monitors and maps that are of use to farmers adopting precision farming in Missouri, United States of America. Firstly, they point out that yield monitors are essential during site-specific nutrient and water management. Yield maps prepared using such monitors provides reliable data regarding soil fertility variations. Subsequently, soil fertility management procedures could be adopted more effectively. It also provides a good feedback regarding influence of an agronomic procedure adopted in a particular management zone or field.

Yield monitors are most useful when loaded on to combines that harvest forage/grain, along with a DGPS receiver. This allows farmers to simultaneously obtain data regarding grain yield at a location, grain moisture content and position data. Yield maps that depict grain productivity variations all across a management zone or field could be prepared using color or shade (Figure 5). It offers excellent data that can be used to supply fertilizer-based nutrients during the succeeding season. Yield monitoring systems that make a periodic report about crop growth or grain yields from different season are also available. The resolution with which, yield monitors record grain yield data is important. For example, a combine harvester collects data at every 2 s. It amounts to 200 data points per acre. However, yield monitors currently used typically provide over 500 points per acre.

According to Casady et al. (2010) there are a few constraints that could reduce accuracy of a yield monitor system. Threshing and grain cleaning steps delay the entry of grains into yield recorders. There is also smoothing of abrupt changes in grain harvest rate. Therefore, both delay and smoothing phenomenon affect accuracy of grain yield recordings. For example, Casady et al. (2010) state that in some combine harvesters/yield monitors, the grain recordings get delayed by 15 s for every 110 ft traversed by the equipment. Yield maps prepared need to take due account of such changes in positioning data. Usually, computer software attached to grain yield monitors do perform precautionary corrections that relate to both delay due to traversing grain and smoothing effect.

In addition to corrections in geopositioning, yield monitors need to account for grain moisture content at a given point. Sensors that monitor both grain yield and moisture simultaneously are often preferred. Instrumentation adopted on combine harvesters may adopt different basic technology to ascertain both grain yield and moisture. Usually, force with which grain hits a recording plate or attenuation of light passing through a stream of grains or grains collected periodically in a cistern or grains passing through a paddle are utilized to estimate grain harvested at a given point in the field. Accuracy of grain yield data, of course depends on speed of the traction equipment, crop species, moisture of grains and so on. Moisture data could help farmers to obtain more accurate knowledge of grain harvests. The combine harvesters equipped with yield monitors are usually provided with 'Yield monitor console' which is a computer/data logger that collects and stores detailed information regarding geo-position, grain yield obtained at each point, grain moisture content at corresponding locations and necessary calibrations. Such spatially indexed data could be effectively used to prepare "Yield maps". Yield maps provide a permanent visible record of productivity at various points in the farm (Figure 5). In addition to grain quantity and moisture content, we can also estimate grain protein content and map it appropriately (Long et al., 2011; Thylen et al., 2002). It is generally believed that grain protein is influenced by the soil N variations. That means over or under fertilization of N does affect grain protein maps. Yet, in most cases, farmers generally use soil fertility maps and yield, while taking decisions regarding crop, fertilizer supply levels and yield goals. Computerized data on soil fertility variation and yield maps could be loaded and used later during variable rate applications (Casady et al., 2010). Printed yield maps prepared using data or yield monitors are most useful to farmers, since they can easily make comparative study of it along with information on soil fertility variations and moisture content of soil (see Figure 5). Yet, we ought to realize that appropriate software, decision support systems, and calibrated fertilizer dispenser systems are needed before yield maps could be effectively utilized during crop production. Over all, accuracy of yield monitors, yield maps and decision support systems depend excessively on the stringent calibrations and accurate data collection systems.

According to Lems et al. (2011), yield monitors supplied by manufacturers often provide accurate data and at best within 1–3% standard error. Irrespective of soil type, crop or region, following basic steps should be adhered in order to achieve best results from yield monitors. They are:

- Proper calibration of the mass flow sensor using multiple loads and Manufacturer's recommendations.
- Inspection of sensors particularly those affected by crop during harvest.
- Verification and calibration of ground speed.
- Correct entry of operating information such as crop type, field, console and so on.
- Proper use of software to extract and process the yield data.

Major components in any yield monitoring system can be depicted as follows (Casady et al., 2010)

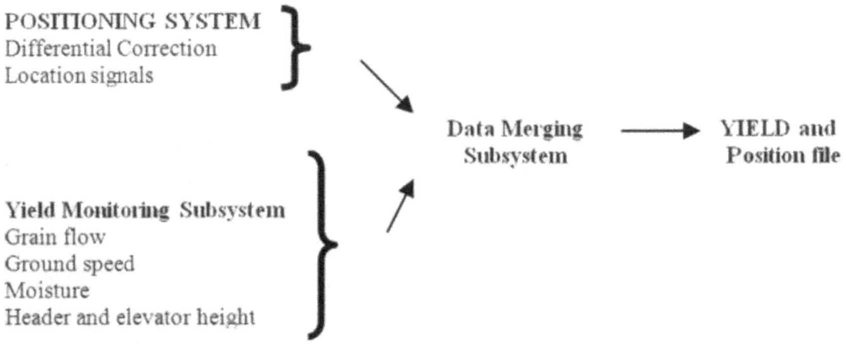

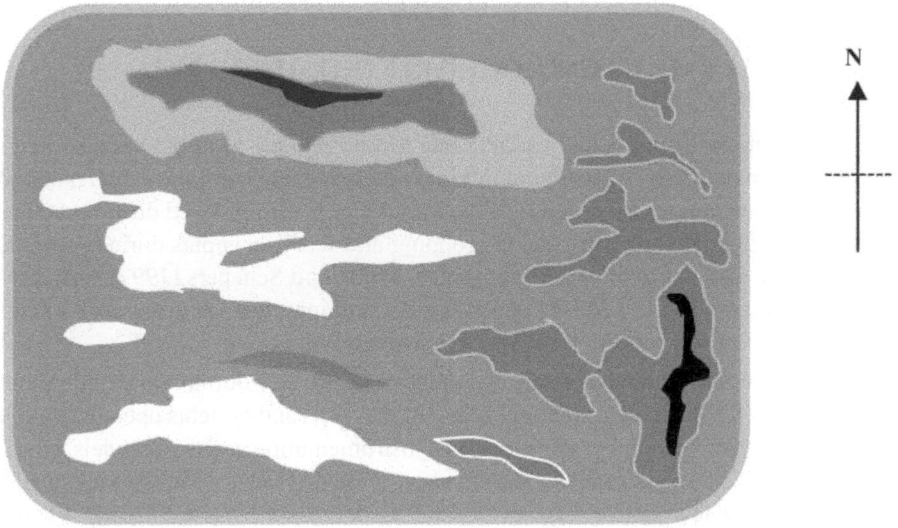

FIGURE 5 A Yield Map of a Soybean field derived using Yield Monitors. Soybean is grown on Haplustalfs in North America. Average grain yield varies between 2 and 3 t ha^{-1}.
Note: Darker color shade indicates greater productivity.

Sensors to Estimate Weed infestation and Density

Weeds affect nutrient dynamics and productivity of crops. They divert nutrients meant for the main crop and compete for several other factors such as soil moisture, photosynthetic radiation, as well as foliage and grain production. Early detection of weed is an important aspect since nutrients supplied under precision farming considers nutrient requirement of maize crop. Opto-electronic sensors that distinguish between green plants have been adopted to trace weeds that emanate within and in between rows. Identification of weeds rapidly using sensors in-between maize rows may actually save 30–40% on herbicide application. Most importantly, timely removal of weeds restricts loss of nutrient to weeds. There are sensors that use spectral signatures of specific weeds that are common to region. Digital cameras placed at vantage points can also monitor fields and highlight occurrence of weeds in between crop rows. Machines with automatic guidance systems and color perception generally distinguish soil, crop, and weeds. They help in eradicating weeds rapidly. Clearly, sensing weed growth via sensors and its eradication has perceptible advantages in terms of nutrient dynamics.

Heisel and Christensen (1999) evaluated a digital camera system that selectively identifies areas with weed infestation in a crop field. Using appropriate color filters, wavelength band reflectance from weed infested zones could be accurately gauged, even when vehicles mounted with sensors traversed at 45 km h^{-1}. Decisions regarding herbicide spray at different locations and at variable rates could be decided based on weed sensor data.

2.4 VARIABLE RATE TECHNOLOGY (VRT): DEFINITIONS AND EXPLANATIONS

Site-specific nutrient supply and variable rate technology are required because fields are not uniform naturally. Agricultural fields are derived from complex parent material that can generate variability in soil fertility. Variability is also induced due to previous management. Spatial variability due to cropping and fertilizer inputs during previous season(s) may turn out to be extensive. Hence, Mulla and Schepers (1997) aptly suggested that potential to use SSNM or VRTs exist and may last for years until a semblance of uniform soil fertility is achieved in a given field.

There are indeed several manuals and guides that provide outlines or details about how to adopt VRT in a given situation. Many of the steps and systems operated within VRT could be specific to geographic location, instrumentation and their models, crops, materials that need to be channeled into fields at variable rates and finally yield goals. Let us consider a few salient features of VRT as described by Giles (2010). Firstly, VRT is a process that involves use of information gathered about soil, its variability and variations in productivity. It also includes making decisions for site-specific agriculture. In the present context, it refers mostly to nutrient management. Essentially, VRT consists of machines and systems for applying a desired rate of crop production materials at a specific time and location. The desired rate at which nutrients or other materials are dispensed could be decided by employing any number of different procedures, computer-based crop models or it could be data accrued through many years of farming a particular zone or plot. Generally, hardware for achieving variable rate supplies of nutrients/water remains same but software and programs that direct the controls in the instrumentations differ depending on context-say crop species, fertilizer formulation, yield goals and so on. Typically, in a farm,

$$\text{Nutrient Application Rate} = \frac{\text{Flow rate of fertilizer (mass or volume)/Unit time}}{\text{Area or land rate covered/Unit time}}$$

Note: "land rate" is the product of width of the machine and ground speed.
Source: Giles, 2010.

In a practical situation, "width of the machine" is fixed but speed of tractor fluctuates and it can be sensed accurately. Therefore, in a VRT or Site-specific nutrient management system, flow rate of fertilizer formulation should be compensated and altered accordingly. The basic task of the VRT is to achieve desired nutrient application rate using computer controlled instrumentation or manually.

A few other terminologies/concepts relevant to VRT are set point application rate, actual or response flow rate, application rate error and spatial resolution. Spatial resolution is defined as smallest area that can receive a distinct and desired application rate. It is the product of lateral and longitudinal resolution. Lateral resolution is the width of each dispenser or fertilizer granules or liquid. Longitudinal resolution is the smallest distance traveled within which application rate is accomplished. The desired application rate that is usually decided by GPS/GIS or sensors and Variable rate application is called the "Set point application rate". The "Set point flow rate" is determined by using set point application rate and the current land rate. The VRT aims at achieving set point flow rate. The flow rate itself depends on minimum response time or dynamic response of the flow control system. We should note that errors creep due to time lapse between set point application rate and actual application rate. Time lapse between signals from computer guided systems and mechanical flow adjusters need due consideration (Figure 6 and Plate 8). Abrupt change in fertilizer supply is difficult to attain under practical farming conditions. However, there are several different in-built and computer guided methods that overcome or compensate for errors in actual computer based decisions and fertilizer supply achieved within a point in the management zone. Over all VRT consists of a "Fertilizer or Nutrient Flow rate regulator system". It is guided by GPS/GIS and a computer-based decision making system that takes a series of algorithms. Any change in application rate at a given time/point involves a certain degree of error based on response time of the instruments and traction equipment. This error could be overcome by adopting specific corrections.

Agricultural crop production techniques vary enormously based on factors related to geographic region, weather pattern, soil type, crops, cropping systems, agronomic procedures and yield goals. Fertilizer supply is one of the most important aspects of crop production in most regions. There are indeed several standardized procedures based on management zones and variable rate applicators to supply fertilizers and other chemicals such as gypsum. Such methods also differ based on cropping zone and region. Let us consider a few examples:

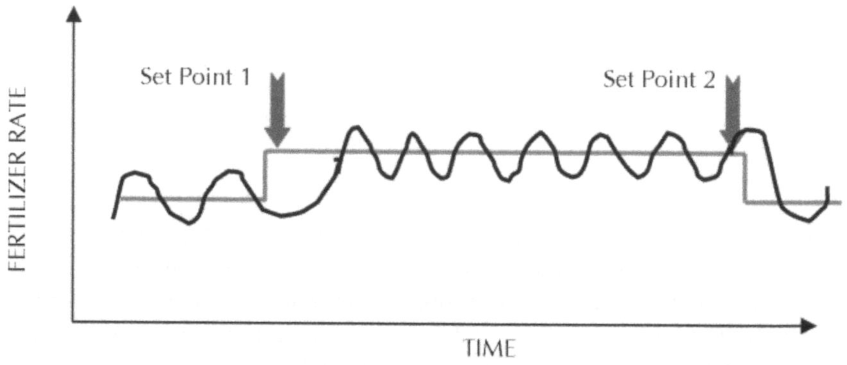

FIGURE 6 Depiction of "Set Point" desired and "Actual flow rate of fertilizer granules" in a "Variable rate Applicator".

Note: It takes time for instruments to react to commands and change the flow rate. Hence, flow rate takes a wavy shape. Usually VRT instruments are mounted on moving vehicles like tractor or trucks. Therefore, at set points, abrupt changes in flow rates (grey straight line) are not possible. Arrows indicate the "set point" where abrupt changes should occur, if VRT were not mounted on a moving tractor or similar vehicle.

Source: Giles 2010

In the Southeast coast of USA soil requires correction for pH. Soil pH varies significantly within a field. Soils may get deteriorated further if lime inputs are either lower or higher than levels needed to correct pH. Hence, lime (gypsum) is supplied accurately using VRT. According to Heiniger and Meijer (2000), use of VRT to apply gypsum to both coastal and tidal zones of Southeast has been profitable.

In the Australian wheat zone, farmers adopt variable rate fertilizer and gypsum supply based on soil type, its fertility status and reaction. Basic steps involved as outline by GRDC (2010) are as follows:

(1) *Set up Management Zones*: Spatial data on soil fertility variations, biomass maps, and grain yield data are used to prepare variable rate maps. As a caution we may note that grain yield/forage maps vary season to season even within the same management zone due to changes in weather pattern. Latest maps are useful. Authentic soil chemical tests, classifications and maps based on availability of nutrients such as N, P, and K is needed prior to directing variable applicators to distribute fertilizers. Create zones based on soil type and moisture status. Commonly, farmers adopt three zones based on inputs required–namely low, medium and high input regions. However, farmers do quite often consider classifying fields into at least 10 zones. This is to help smooth changeover of variable rate applicators as they change input rates.

(2) *Determine Fertilizer Input Rates*: Fertilizer supply in most cases is dependent on crop response to nutrient supplied and economic advantages. It is also dependent on amount of nutrients depleted in the previous season and expected recovery by the crop sown. Consulting fertilizer response curves relevant to soil type, crop and region are highly recommended. Fertilizer supply, especially N could be split to match crop requirements at different growth stages. Nutrient replacement approach usually depends on extensive soil test and plant analysis data from different management zones. A few other techniques to supply fertilizer-based nutrients depend on yield goal and economic considerations. Computer-based simulations are being currently used to evaluate various fertilizer dosages and supply schedules. In fact, quite often simulation data or inferences are fed to variable rate applicators, so that nutrient supply is accurate.

(3) *Producing fertilizer Prescription Maps*: Mapping software that prescribe input rate at each management zone is prepared. Data files are loaded into computers that control variable rate applicators. Unfamiliar software should be avoided.

(4) *Conduct a Test Trip*: Conducting a test trip to optimize instruments and fine tune input rate is required. A dry run to test the prescription maps and functioning of variable rate applicators is most important.

(5) *Soil and Crop tests*: Soil and crop tests at each of the strips or management zones could be used to confirm accurate functioning of instruments and computer software meant for precision farming.

Adoption of Variable Rate Technology in Different Countries

According to Lowenberg-DeBoer (2003a) fertilizer application variable rate technology (VRT) is a high profile agronomic procedure adopted in the Great Plains and other intensive farming zones of North America. In 1996, nationwide 29% of all retail farms in USA were using VRT to spread fertilizers and it was based on prior knowledge about soil fertility variations based on grid samples and fertility maps. Adoption of VRT spread rapidly since then. In 2002, 50% of farms made use of controlled VRT for fertilizer supply. The VRTs became more frequent in the Midwest where corn production dominated (Plate 9 and 10). The VRTs were also used frequently on other crops like sugarbeet and potatoes. In 2003, about 10% of potatoes and 9% sugarbeet area in the Idaho/North Dakota region was being supplied with fertilizers using VRT. The use of VRT for other purposes like pesticide application is scattered in USA and other regions. Surveys in 1999 suggested that only 1–3% cereal farms in North America used VRT to spray pesticides. However, in due course, use of VRT has improved considerably.

PLATE 8 Variable rate control system used to dispense exact quantity of Fertilizer-N at each spot in the field under prescriptive agriculture (Precision Agriculture).

Note: The small motor fitted by the side of each row is regulated by signals from Decision support system. It regulates fertilizer-N flow based on pre-determined maps or "on-the-go" signals from decision support systems. Signal transmission itself may take time to reach. Abrupt changes in the speed and fertilizer flow may not occur because tractors are always on the move. Therefore, calibrations are needed to achieve greater accuracy in fertilizer dispenser.

Source: Mr. David Nelson, Nelson Farms Inc, Fort Dodge, Iowa, USA.

Soil doctor is a variable application system based on soil NO_3-N estimates derived using sensors. According to Colburn (1999), performance of Soil doctor is actually dependent on extent of variability encountered with regard to soil N, crop response pattern to N inputs and inherent soil N levels. The high calibration level in the instruments makes VRT to supply more N in patches that are deficient. Yet, fertilizer-N supplied using sensor-VRT system (Soil doctor) is much less compared to conventional system. Use of Soil Doctor also reduced soil N loss *via* leaching, percolation, and erosion (see Plate 9 and 10).

PLATE 9 Top: Variable rate nitrogen and phosphorus applicators used to dispense basal nutrients into a field to be sown with Maize in a farm located near Fort Dodge, Iowa, USA.

Bottom: A close-up view of intricate tubing and connections that make up a variable rate fertilizer applicator.

Note: The variable rate inputs are guided from fertilizer N and P containers through elaborate tubing and injectors at each spot within a row. It helps in spreading fertilizer N and P in exact quantities as stipulated by soil nutrient maps or "on-the-go" decision support system.

Source: Mr. David Nelson, Nelson Farms Inc, Fort Dodge, Iowa, USA.

PLATE 10 Top: A Variable rate N applicator accomplishing Split N supply on maize seedlings planted on Haplustalfs of Northern Iowa.

Bottom: A view of the field after Split application of fertilizer-N using variable arte technique.

Note: Individual applicators aligned with each row dispense quantities of Split N as directed by the decision support system.

Source: Mr. David Nelson, Nelson farms Inc, Fort Dodge, Iowa USA.

Bongiavanni and Lowenberg-DeBoer (2005) believe that spread of VRTs in Argentina would be slow yet steady. This is due to high cost of soil sampling, low management induced variations in soil fertility and current low rates of fertilizer inputs practiced by farmers.

2.5 SOIL FERTILITY MAPS

Quite often, farmers adopting precision farming techniques prepare a nutrient map of the field. It practically describes soil fertility variations and provides a rough guide while selecting suitable cropping pattern and agronomic procedures. Nutrient maps provide better understanding of the soil resources, so that farmers can suitably match it with crop species. A nutrient map is a good basis to improve grain/forage yield, reduce on fertilizers and manage nutrient dynamics appropriately. A nutrient map helps farmers in noting soil pH and nutrient variations, so that they may choose appropriate crop species to suit each field. It allows farmers to identify areas that need relatively less fertilizer, so that he can save on fertilizer cost. Farmers can balance fertilizer supply with nutrient removal rates of different crops grown. Nutrient maps prepared periodically provide farmers with an idea about the extent of soil degradation, loss or accumulation of nutrients and risks involved. Let us consider an example. Paddy cultivation is wide spread in Southeast Asian wetlands. Obtaining uniformly optimum or high soil fertility is indeed a major constraint. Therefore, farmers intending to adopt precision farming approaches often prepare a soil map depicting the variations in soil fertility, especially major nutrients, pH, and organic matter.

Aishah et al. (2010) state that in countries such as Malaysia, paddy cultivators select georeferenced soil samples and assess them for major, secondary and micronutrients. In addition they also keep track of soil pH and organic matter content. Soil maps showing these physico-chemical traits are then prepared and utilized during variable rate inputs. Kriged maps have also been used routinely. The accuracy of soil mapping and its usefulness regarding resolution of soil fertility variations depends on sampling depth(s), distance between grid points and timing.

Aspect	Map-based	Sensor-based
Methodology	Grid sampling, Lab analysis, Soil maps, Variable-rate applicators	Real time Sensors, immediate feed back, computer-based decision support, variable rate applicators
GPS/DGPS	Required	Not necessary
Laboratory analysis	Required	May not be required
Mapping	Required	May not be needed
Time consumption	Relatively More	Relatively Less

Aspect	Map-based	Sensor-based
Limitations	Cost of Soil and Plant testing could limit its use	Lack of apt sensors to monitor soil/plant traits and information on soil could hinder operations
Operation	Difficult	Easy
Skilled Personnel	Required	Required
Sampling Unit	2 to 3 acres	Individual spot
Relevance	Popular in developing countries	More frequently used by farmers in Developed countries

Source: Patil and Shanwad, 2010

Crop Growth and Grain Yield Maps

Yield mapping is one of the first steps in implementing precision farming. Yield maps are generally produced rapidly by processing data collected on the combine harvesters that has a vehicle positioning system integrated with grain/forge yield recording system. Earliest combines with ability to process yield data and provide yield map, were produced by Massey Ferguson Inc in USA. Such combine harvesters usually have a DGPS fitted. Generally, output from the combine harvester consists of electronically stored data files. Data points recorded every 1.2 s along with longitude and latitude could be processed (kriged) and detailed crop yield map generated. The yield maps are smoothened in order to get a better idea of the yield variations. The need for adoption of precision farming is dependent on the extent of variation shown in the yield maps retrieved from combines (Blackmore, 2003). Precision techniques may not be needed if variations in yield are not significant. During practical farming, maps should be properly processed in order to be able to compare them appropriately. For example, maps of a field from different years. There are different software's (e.g. Surfer) that decode, interpret and prepare accurate maps depending on purpose (Blackmore 2003).

Maps for Precision Farming: Overview

Firstly, let us consider an outline of the complete set of procedures that go to accomplish precision farming in any region. As stated earlier in Chapter 1 there are two major approaches. One of them is map based and the other is real time sensor based. In the map-based system previous data is extensively used to build maps that depict yield, soil variations, pestilence, soil moisture distribution, and so on. Based on grain/forage yield goals and economic constraints, decision support systems that utilize computer models/simulations help the farmer in preparing maps and software that allow him to supply seeds, fertilizers, irrigation, pesticides and other amendments, exactly to match, and overcome the variation.

Mapping Software	**Precision function accomplished**
Yield Maps	Grain processing and Yield recording; Forage harvest and Yield recording
Soil Nutrient Maps	These are prepared using Soil chemical test data, They depict soil fertility variations, and help in preparing soil nutrient and fertilizer distribution maps that includes parameters like pH, salinity, Al toxicity
Application Maps	Based on specific computer-amenable models and accrued data, decision support systems prepare a set of soft ware or maps that help farmer to accomplish various tasks accurately: Such maps help in deciding tillage intensity, sowing depth, seeding density, spacing, fertilizer inputs, depth of fertilizer placement, soil moisture supply, pest control sprays and so on.

The second approach called "real time approach" involves rapid on-the-go analysis of soil and/or crop characteristic, computer-based processing of data and prescription in a matter of seconds. The commands generated are used to accomplish specific tasks like tillage, seeding, fertilizer supply, irrigation, pesticide spray, and so on.

Data Accrual (Geo-referenced or on-the-go)	**Precision function accomplished**
Yield Maps	Harvesting and recording grain/forage quantities
Soil Nutrient Maps	Rapid sensor-based soil analysis and developing suitable Electronic signals/commands to supply exact quantity of fertilizers or other amendments
Application Maps	Seeding density, fertilizer supply, irrigation

Source: Based on depictions by Wagner, 1999.

KEYWORDS

- **Agronomic procedures**
- **Electrical conductivity**
- **Electromagnetic Induction**
- **Grid sampling**
- **Harvesters**
- **Infrared thermography**
- **Soil organic matter**

REFERENCES

Aimrun, W., Amin, M. S. M., Amin H., and Nouri, H. Paddy field zone characterization using Apparent Electrical Conductivity for Rice Precision Farming. *International Journal of Agricultural Research*, **6,** 10–28 (2011).

Aishah, A. W., Zauyah, S., Anaur, A. R., and Fauziah, C. I. Spatial variability of selected Chemical characteristics of Paddy soils in Sawah Sempadan, Selangor, Malaysia. *Malaysian Journal of Soil Science*, 14, 27–39 (2010).

Atherton, B. C., Morgan, M. T., Shearer, S. A., Stombaugh, T. S., and Ward, A. D. Site-Specific Farming: A perspective on information Needs, Benefits, and Limitations. *Journal of Water and Soil Conservation*, **54**, 455–461 (1999).

Balkcom, K. and Rodekohr, D. Variable Rate Cotton fertilization Development in the Coastal Plains of Alabama. American Society of Agronomy Annual meetings Long Beach CA, USA, pp. 1–2 (2010), Retrieved from ASA_10_greenseeker.pdf (January 1st, 2011).

Blackmore, S. *The role of Yield maps in Precision Farming.* National Soil Resources Institute, Cranefield University, Silsoe, United Kingdom (Doctoral Dissertation), pp. 161 (2003).

Bogrecki, I. and Lee, W. S. Comparison of Ultraviolet, Visible and Near Infrared Sensing for Soil Phosphorus. *Biosystems Engineering*, **96**, 293–299 (2007).

Bongiovanni, R., and Lowenberg-DeBoer, J. *Precision Agriculture in Argentina.* Third Simposio Internacional de Agricultura de Precision, pp. 1–14. (2005), Retrieved from http://www.cnpms. embrapa.br/siap2005/palestras/SIAP3-Palestra_Bongiovanni_e_LDB.pdf (March 25th, 2011).

Bredemeier, C. and Schmidhalter, U. Non-contacting Chlorophyll fluorescence sensing for Site-specific Nitrogen fertilization in Wheat and Maize. In *Precision Agriculture*. J. Stafford and A. Werner (Eds.). Wageningen Academic Publishers, pp. 103–108 (2003).

Campbell, R. H. Recent Advances in Yield monitoring of conveyor harvested crops. In *Proceedings of the 4th International Conference on Precision Agriculture*. P. C. Roberts (Ed.). American Society of Agronomy, Madison, WI, USA, pp. 1101–1106 (1998).

Casady, W., Pfost, D., Ellis, C., and Shannon, K. *Precision Agriculture: Yield Monitors*. University of Missouri System, University Outreach and Extension Service. *Water Quality*, **451**, 1–4 (2010).

Christy, C. D., Drummond, P., and Lund, E. Precision Agriculture applications of an 'on-the-go' soil infra-red reflectance sensor, pp. 1–12 (2010), Retrieved from http://www.veristech-com/pdf_files/ Optical_8thinticonf.pdf (June 12th, 2011).

Chung, S., Sudduth, K., Jung, Y., Hong, Y., and Jung, K. Estimation of Korean Paddy field soil properties suing Optical reflectance. In *ASABE Annual International Meeting Technical papers*. American Society of Biological Engineers Annual International Meeting. Providence, Rhode Island. Paper No. 083682, pp. 1–3. (2008), Retrieved from http://asae.frymulti.com/abc.asp?JID=5&AID=25021&CID=prov2008&T=2 (December 15th, 2010).

Colburn, J. W. *Soil Doctor a multi-parameter, real time Soil sensor and concurrent input control system*. Proceedings of the 4th International Conference on Precision Agriculture, p. 1693 (1999), Retrieved from linkinghub.elsevier.com/retrieve/pii/S001670610100074X (May 28th, 2011).

Dobermann, A., Blackmore, S., Cook, S. E., and Adamchuk, V. I. Precision Farming: Challenges and Future Directions. In *New Directions for a Diverse Planet*. Proceedings of the Fourth International Crop Science Congress, Brisbane, Australia, p. 19 (2004), Retrieved from www.cropscience.org.au (January 20th, 2011).

Doerge, T. Yield monitors create on and off farm profit opportunities. *Crop Insights, Pioneer International*, 9, 1–4 (1999).

Doerge, T., Kitchen, N. R., and Lund, E. D. Soil Electrical Conductivity mapping. International Plant Nutrition Institute. Norcross, Georgia, USA, pp. 1–2. (2011), Retrieved from http://www. ipni.net/ppiweb/ppibse.nsf/$webindex/article=BD1CF45C852569D700636EDAC9ADC4DE (June 27, 2011).

Ferguson, R. W. and Hergert, G. W. Soil sampling for Precision Agriculture. *Journal of Animal and Plant Sciences*, **5**, 494–506 (2009).

Franzen, D. W. and Kitchen, N. R. *Developing Management zones to target Nitrogen applications*. International Plant Nutrition Institute, Norcross, GA, USA, pp. 1–2 (2011), Retrieved from http://www.ipni.net/ppiweb/ppibse.nsf/$webindex/article=772BE&3B8525695A005A12E99F5E03CD (June 27th, 2011).

Franzen, D. W. and Swenson, L. J. Soil sampling for Precision farming. *Sugar beet Research and Extension Reports*, **26**, 129–134 (1995).

Geypens, M., Vanaongeval, L., Vogels, N., and Meykens, J. Spatial variability of Agricultural Soil fertility parameters in a Gleyic Podzol. *Precision Agriculture*, **1**, 319–326 (1999).

Gholizadeh, A., Amin, M. S. M., Anuar, A. R., and Aimrun, W. Evaluation of SPAD-Chlorophyll meter in two different Rice growth stages and its Temporaltemporal variability. *European Journal of Scientific Research*, **37**, 591–598 (2009).

Giles, K. *Variable Rate Technology (VRT) for Site-Specific Agriculture*, pp. 1–7 (2010), Retrieved from. Precision ag.org/word/ch12.doc (January 11th, 2011).

Grain Research and Development Corporation (GRDC). *Precision Agriculture Fact sheet*. How to put Precision Agriculture into practice. Grain Research and Development Corporation, Kingston, Australia, pp. 1–6 (2010), Retrieved from www.grdc.com.au (January 1st, 2011)

Hall, T. L., Backer, L. F., Hoffman, V. L., and Smith, L. J. Evaluation of Sugar beet Yield sensing systems operating concurrently on a harvester. In *Proceedings of the 4th International Conference on Precision Agriculture*. P. C. Roberts (Ed.). American Society of Agronomy, Madison, WI, USA, pp. 1107–1118 (1998).

Heisel, T. and Christensen, S. A *Digital camera system for Weed detection*. Proceedings of the 4th International Conference on Precision Agriculture. American Society of Agronomy, Madison, USA, pp. 1569–1577 (1999).

Hellebrand, H. J. and Umeda, M. *Soil and Plant Analysis for Precision Agriculture*. 1st Asian Conference on Precision Agriculture, Kuala Lumpur Malaysia, pp. 1–8 (2004), Retrieved from http://www.atb.potsdam.de/hauptse-deutsch/institute/abteilungen/abt2/Mitarbeiter/jhellebrand/jhellbrand/publikat/sensing.pdf (February 21st, 2011).

Huang, C. W., Huang, C. C., Yang, C. K., Wu, T. H., Tsai, Y. Z., and Miao, P. L. Determination of Nitrogen content in Rice crop using Multispectral Imaging. *ASAE Annual Meeting*. Paper No. 031132, pp. 1–2 (2003) http://asae.frymulti.com/ abstract.asp=13741&t=2http://asae.frymulti.com/ abstract.asp=13741&t=2 (January 11th, 2011).

Jahanshiri, E. *GIS-based Soil sampling methods for Precision Farming of Rice*. MastersMaster's thesis, University Putra Malaysia, Selangor, Serdang, Malaysia, p. 72. (2006), Retrieved from http:// psasir.upm.edu.my/612 (April 23rd, 2011).

Johnston, A. E., Barraclough, P. D., Poulton, P. R., and Dawson, C. J. Asessment of some spatially variable soil factors limiting crop yields. In *Proceedings of International Fertilizer Society*, pp. 1–2 (1998), Retrieved from http://www.fertilizer-society.org/proceedings /uk/Pro419.htm (June 12th 2011).

Jorgensen, R. N. *Study on Line-imaging Spectroscopy as a tool for Nitrogen Diagnostics in Precision farming*. Department of Agricultural Sciences, Riso National Laboratory, Roskilde, Denmark, PhD Thesis, p. 93 (2002).

Kasowski, M. and Genereax, D. *Farming by the foot in the Red River valley of Minnesota. Agricultural Finance*, **124**, 20–21 (1994).

Kebabian, P. L., Theisen, A. F., Kallelis, S., Scott, H. E., and Freedman, A. A passive two-band plant fluorescence sensor with applications in Precision Agriculture. *Precision Agriculture and Biological Quality*. Proceedings of the SPIE, Boston, MA, USA, pp. 238–245 (1998).

Khosla, R. *Active Remote Sensing: An innovative Technique for Precision Fertilization*, pp. 1–4 (2008), Retrieved from http://www.extsoilcrop.colostate.edu/newsletter.htm (March 1st, 2011).

Kim, J. H., Sudduth, K. A., and Hummels, J. W. Soil Macronutrient Sensing for Precision Agriculture. *Journal of Environmental Monitoring*, **11**, 1810–1824 (2009).

Kyveryga, P. M., Tao, H., Morris, T. F., and Blackmer, T. M. Identification of Nitrogen Management categories by Corn stalk Nitrate sampling guided by Aerial Imagery. *Agronomy Journal*, **102,** 858–867 (2009).

Kyveryga, P., Blackmer, T., and Pearson, R. Uncaliberated late-season Digital imagery for predicting Corn Nitrogen status within fields. International *Annual Meetings of American Society of Agronomy*., Long Beach, California, USA, 107(3), p. 1 (2010a), Retrieved from http;//a-c-s.confex/com/crops/2010am/webprogram/paper58309.html (January 23rd, 2011).

La, W., Sudduth, K. A., Chung, S., and Kim, H. Spectral reflectance estimates of Surface soil Physical and Chemcial properties. In *ASABE Annual Interntional Meeting, Technical papers*. American Society of Agricultural and Biological Engineers Annual International meeting Providence, Rhode Island. Paper No. 084242, pp. 1–3 (2008), Retrieved from http://asae.frymulti.com/abc.asp?JID=5&AID=24696&CID=prov 2008&T=2 (December 15th, 2010).

Lakes, J. V., Bock, G. R., Goode, J. A., and Mulla, D. J. *Geostatics, Remote Sensing and Precision Farming*. Proceedings of CIBA foundation symposium on Precision Agriculture, pp. 1–2 (2007). Retrieved from http://onlinelibrary.wiley.com/doi/10.1002/9780470515419.ch7/summary;js.htm (July 12th, 2010).

Lee, K. S., Lee, D. H., Sudduth, K., Chung, S., and Drummond, S. Wavelength Identification for Reflectence estimation of Surface and Subsurface soil properties. *ASABE Annual Interntional Meeting Technical papers*, pp. 1–2. (2007), Retrieved from http://asae.frymulti.com/request.asp?search=1&JID=5&AID=22894&CID=min2007&=&i=&t=2 (March 1st, 2011).

Lemos, S. G., Nogeira, A. A., Torre-neto, A., Parra, A., and Alonso, J. Soil Calcium and pH monitoring sensor system. *Journal of Food and Agricultural Chemistry*, **55**, 4658–4663 (2007).

Lems, J., Clay, D. E., Hamburg, D., Doerge, T. A., and Resse, S. C. *Yield Monitors-Bsic steps to ensure system accuracy and performance*. International Plant Nutrition Institute, Norcross, Georia, USA, pp. 1–2 (2011), Retrieved from http://www.ipni.netppiweb/ppibse.nsf/$webindex/article=4E42F3D2852569C4006C0399A8BB658B (June 27th, 20111).

Long, D. S., Engel, R. E., and Reep, R. *Grain protein sensing to identify Nitroegen management of Spring wheat*. International Plant Nutrition Institute, Norcross, Gerogia, USA, pp. 1–2 (2011).

Lowenberg-Deboer, J. *Precision Framing or Convenience Farming*, pp. 1–32. (2003a), Retrieved from http://www. regional.org.au/au/asa/2003/i/6/lowenberg.htm (March 23rd, 2011).

Maleki, M. R. *Soil-sensor–based Variable rate Phosphorus fertilization: 'On-the–Go' fertilization*. Lambert Academic Publishing, London, p. 244 (2010).

Mallarino, A. P. Using Precision Agriculture to improve Soil fertility Management and On-farm Research. Iowa State University Extension Services. *Integrated Crop Management*, **480**, 12–14 (1998).

Moshia, M. S., Khosla, R., Davis, J. G. Westfall, D., and Reich, R. *Precision Manure Management across Site-specific Management Zones*. International Annual Meetings of American Society of Agronomy. Long Beach, California, USA, pp. 102–1:1–2 (January 11th, 2011) (2010).

Moulin, A., Derksen, D., Mclaren, D., and Grant, C. *Spatial variability of soil fertility and identification of management zones on hummocky terrain*. Brandon Research Center, agriculture and Agri-Food, Canada. Research Report, pp. 1–8 (2003).

Mulla, D and Schepers, J. Key processes and properties for Site-specific Soil and Crop Management. In *The Site of Site-specific management for Agriculture*. F. J. Pierce and E. J. Sadler (Eds.). American society of Agronomy, Madison, WI, USA, pp. 1–18 (1997).

Patil, V. C. and Shanwad, U. K. *Relevance of Precision Farming to Indian Agriculture*. Department of Agronomy, University of Agricultural Sciences, at Raichur, Karnataka, India, (2010), Retrieved from www.acr.edu.in/info/infofile/144.pdf (May 28th, 2011).

Peng S., Garcia, F. V., Laza, R. C., Sanico, A. L., Visperas, R. M., and Cassman, K. G. Increased N-use efficiency using Chlorophyll meter on High yielding Irrigated rice. *Field Crops Research*, **47,** 243–252 (1996).

Peng, S., Laza, R. C., Garcia, F. C., and Cassman, K. R. Chlorophyll meter estimates Leaf area-based N concentration of Rice. *Communication of Soil and Plant Analysis*, **26,** 927–935 (1995).

Plattner, C. E. Hummel, J. W., Robert, P. C. Rust, R. H., and Larson, W. E. *Corn plant population sensor for Precision Farming Agriculture*. Proceedings of the 3rd International Conference in Precision Agriculture. Minnesota, USA, pp. 785–794 (1996).

Rossel, R. V., Walter, C., and Fouad, Y. Assessment of two reflectance techniques for the quantification of the within-field Spatial variability of Soil Organic Carbon. In *Precision Agriculture*. J. Stafford, and A. Werner (Eds.). Wageningen Academic Publishers. Netherlands, pp. 697–703 (2003).

Schmidt, J. D., Dellinger, A. E., and Beegle, D. B. Nitrogen Recommendations for Corn: An on the Go sensor compared with current recommendation methods. *Agronomy Journal*, **101,** 916–924 (2009).

Selige, T., Natscher, L., and Schmidthalter, U. Spatial detection of Topsoil properties using Hyper-spectral sensing. In *Precision Agriculture*. J. Stafford and A. Werner (Eds.). Wageningen Academic Publishers, Netherlands, pp. 633–638. (2003).

Shaver, T., McCuen, R. H., Ferguson, R., and Shanahan, J. *Crop canopy Sensor utilization for Nitrogen management in Corn under semi-arid limited irrigation conditions*. American Society of Agronomy Annual International Meetings, Long Beach California, USA (Abstract), p. 1 (2010).

Snyder, C. S., Bruulsema, T. W., Sharpley, A. N., and Beegle, D. B. *Site-specific use of the environmental phosphorus concept*. International Plant Nutrition Institute, Norcross, Georgia, USA, pp. 1–2 (2011), Retrieved from http://www.ipni.net/ppiweb/ppibase.nsf/ $webindex/article=28F949238525695300581E031A2C31B0 (June 27, 2011).

Solari, F., Shanahan, J. H., Ferguson, R. B., and Adamchuk, V. I. An active sensor algorithm for corn N applications based on chlorophyll meter supply index frame work. *Agronomy Journal*, **102**, 1090–1098 (2008).

Sonka, S. T., Baeur, M. E., and Cherry, E. T. *Precision Agriculture in the 21st Century: Geospatial and Information Technologies in Crop Management*. National Academy Press, Washington D.C., p. 168 (1998).

Stroppiana, D., Boshetti, M., Alessandro Brivio, P., and Bochi, S. Plant Nitrogen concentration in Paddy rice from Field Canopy Spectral Radiometry. *Field Crops Research*, **11,** 119–129 (2009).

Sudduth, K. A., Hummel, J. W., and Funk, R. C. *Soil Organic Matter sensing for Precision herbicide application*. Proceedings of Conference on Pesticide formulations and application systems: Tenth symposium Philadelphia, USA, pp. 111–125 (1990).

Sudduth, K. A., Kitchen, N. R., Scharf, P., Palm, C., and Shannon, H. *Field-scale N application using Crop Reflectance Sensors*. American Society of Agronomy Annual Meetings Abstracts New Orleans, USA, Paper No 153–157, p. 1 (2007).

Sudduth, K., Newell, K., and Scott, D. *Comparison of three Canopy Reflectance Sensors for Variable-rate Nitrogen application in Corn*. Proceedings of the International Conference on the Precision Agriculture Abstract, pp. 1–2 (2010), Retrieved from www.ars.usda.gov/ pandp/people/people.htm?personid=1471.html (December 15th, 2010).

Thylen, L., Gilbertsson, M., Rosenthal, T., and Wren, S. An on-line protein sensor-from research to product, pp. 1–9. (2002), Retrieved from http://www.zeltex.com/online_proteinsensor.doc (June 11th, 2011).

Upadhyaya, S. and Teixeira, A. Sensors for Information gathering during precision farming, pp. 1–9 (2010), Retrieved from www.docstoc.com/docs/34311781/Sensors-for-Information-Gathering (January 4th, 2011).

Wagner, P. *The Future of Precision Farming*. The Development of a Precision Farming information system and Economic aspects, pp. 1–13 (1999), Retrieved from. http://www.lb.landw.uni-halle.de/publikationen/pf/pf_efita99.htm (June 2, 2011).

Zhang, J. H., Wang, K., Bailey, J. S., and Wang, R. C. Predicting Nitrogen status of Rice using Multi-spectral Data at Canopy scale. *Pedosphere*, **16**, 108–117 (2006).

3 Precision Farming, Soil Nutrient Dynamics, and Crop Productivity

CONTENTS

3.1 PRECISION FARMING INFLUENCES VARIOUS ASPECTS OF NUTRIENT DYNAMICS

Precision farming influences basic processes that have immediate effect on nutrient dynamics within a field or cropping zone. Nutrient dynamics within a field, farm or cropping expanse or even a large agroecosystem is basically affected by series of interacting processes in soil, crop phase, and atmosphere. The intensity, timing, and extent to which a particular process gets affected are important. In any agroecosystem, most commonly known processes that are relevant to maintenance of nutrient dynamics are nutrient supply that is extent to which a field or farm is impinged with extraneous nutrients. It may be *via* inorganic or organic fertilizers or recycling of biomass or even natural deposits. Agronomic aspects like precise placement and distribution of manure in soil also need due attention. Precision farming procedures may also affect nutrient recovery rates, nutrient loss *via* natural process like percolation, seepage, erosion, runoff, emission, and volatilization. Precision farming primarily stipulates placement of exact quantity of fertilizer and chemical amendments in the profile and it is based on crop's demand at various stages. Such precise supply and timing of fertilizers may influence the physico-chemical nature of soil in a way different to that seen under conventional systems. Long-term effect of precision farming also needs due attention. Most importantly, almost all of the soil nutrients that crops require undergo series of chemical transformations, at different rates. Nutrient transformations may be marginally or severely influenced by precision techniques. This has to be quantified and taken into account within computer models used for decision support systems. For example, N mineralization may add appreciable amount of N into available N pool that needs to be considered during variable rate N supply. The extent of residual nutrients carried further to next season, accumulation pattern and localization are equally important. Precision farming does affect residual nutrient in soil. It limits this fraction. Residual nutrient has to be quantified and considered during variable rate supply. Since nutrient accumulation is meager under precision farming, nutrient loading into drains, channels, and ground water gets minimized. This is perhaps among the most important environmental benefits derived due to supply of accurate amounts of fertilizers. Biomass and mineral nutrients recycled in a field or agroecosystem could be immensely influenced by the precision techniques.

3.1.1 Precision Farming and Nutrient Supply

Nutrient inputs prescribed under precision farming, depend generally on soil fertility variation, yield goal and economic benefit. Nutrient supply and its timing within a crop season is perhaps the first process that precision techniques affect. Incidentally, nutrient supply is among the most important aspects of nutrient dynamics within a field or agroecosystem. It has far reaching consequences on a series of other soil and crop related process like nutrient availability, nutrient absorption and its transfer to crop phase, nutrient transformation rates, nutrient accumulation pattern, percolation seepage, emission and so on. Fertilizers and amendments supplied under precision techniques are actually guided by computer-based decision support system. Computer models that consider soil processes, crop growth rates, and nutrient demand and yield goals are utilized to arriving at most appropriate dosage for each spot in a field. This is unlike uniform rates prescribed for large expanses, districts or states by the governmental agencies. During a precision farming exercise, nutrients are usually supplied exactly and in quantities as needed by the crop at various strategies. Fertilizer-based nutrient and Farmyard Manure (FYM) prescribed under precision techniques duly considers several soil related aspects like physico-chemical properties, soil nutrient transformation, nutrient availability, root growth and nutrient absorption rates, nutrient depletion rates, and most importantly crop species, its genotype and yield goals. Further, fertilizers prescribe under precision technique is generally trifle lower than under farmer's practice. Therefore, it leaves small or no residual nutrients in soil profile. Fertilizer is placed exactly near the roots. Hence, nutrients are efficiently absorbed. Most importantly, since nutrient distribution is based on prior knowledge of variations that occur in individual fields, immediately after a fertilizer supply exercise, soil nutrients are held uniform throughout the field. This process may have direct consequences on nutrient stratification, microbial population and its distribution, and nutrient transformations. Precision techniques provide due consideration to nutrient ratios and balance in soil. Overall, reduction in nutrient supply to a single field kept under precision farming technique or if extrapolated to entire cropping zones is conspicuous. It has immediate effect on fertilizer use efficiency. It also has far reaching effects on environmental parameters, crop yield, and economic advantages.

3.1.2 Precision Farming and Nutrient Recovery

Soil nutrient recovery by the crop is another basic process related to nutrient dynamics in a field or an ecosystem at large. Since precision techniques remove variations in soil nutrient availability within a field, it could be easily anticipated that nutrient removal rates in the entire field would be uniform. Whereas, under conventional farming systems, nutrient recovery rates are variable. In other words, nutrient shift from soil to crop phase is more uniform under precision farming. Obviously, aspects like root growth rate, nutrient absorption and accumulation rate fluctuate less within a field kept under precision technique. Nutrient recovery from precision fields could be more efficient since nutrient placement, timing and distribution in the soil profile is exact and predetermined to match crop's demand. In a conventional farm, nutrient recovery rates may get affected adversely during later stages because of erratic depletion rates, loss and exhaustion by the time crop reaches mid or later stages of growth. This is avoided in a precision technique, since nutrient supply is spaced, split and applied at quantities as required at each stage of crop. Traditional farming procedures may underestimate nutrient needs for the entire season and result in deficiency in the soil process at crucial stages of growth. This clearly reduces yield formation. Such an error is removed under precision technique, since nutrient needs are met exactly at any stage of the crop. Most importantly, in a field or even in an agroecosystem, nutrient recovery from soil through roots and on to shoots/grains is an important phenomenon that has direct consequences on nutrients held in the various portions of field or an agroecosystem. Quantum of residual nutrient held in the soil profile is affected, based on percentage nutrients recovered from fertilizers. Nutrients vulnerable to loss *via* percolation, seepage, surface flow or emissions would be proportionately low, if recovery is efficient, as it happens to be under precision farming.

3.1.3 Precision Farming and Nutrient Loss

Precision farming is spreading rapidly in many agricultural regions. Researchers hope that it helps in stabilizing ecosystems by reducing loss of nutrients from agro-zones. Precision farming actually helps farmers to reduce fertilizer and pesticide supply by making the process more accurate. Farmers can supply fertilizers with great accuracy and in quantities needed exactly by the crop in question. Supply of exact quantities means efficient use without undue accumulation of nutrients in soil profile. Otherwise, accumulated nutrients could become vulnerable to loss *via* leaching, percolation or runoff. Such nutrient loss could be avoided if precision farming is practiced. Splitting fertilizers and matching nutrient need based on each crop stage, firstly avoids persistence of nutrients in the soil profile. Nutrient movement away from root zone, either laterally or vertically down in to vadose zone, gets reduced immensely. Since fertilizers are placed exactly nearer to root zone, it is scavenged efficiently and quickly without leaving residual fraction that could be vulnerable to emissions. Nutrient loss *via* emission is reduced under precision farming.

Precision Farming and Nutrient Use Efficiency

Nutrient use efficiency is an important aspect in any field, individual farm or even a vast agroecosystem. It has direct relevance to nutrient supply, recovery rates, accumulation pattern in soil and crop biomass, and residual effect and recycling patterns. Of course, it affects crop productivity. Precision farming almost always involves supply of exact quantities of nutrients in time and space. It also envisages most appropriate placement of split dosages nearer the root zone. Precision farming provides nutrients at a uniform rate; hence, undue dilution or accumulation of nutrients in plant tissue is avoided. It means nutrient use efficiency is held constant within the crop phase of the ecosystem. Precision farming also maintains nutrient balance and appropriate ratios of nutrients, so that deficiency of any single element does not affect normal recovery and use of major nutrients. Hence, most often we derive comparatively higher nutrient use efficiency by adopting precision techniques. Precision techniques that are guided by computer models, decision support systems, and variable rate inputs literally aim at better fertilizer efficiency.

Major nutrients that are generally needed/supplied in higher quantities into a field have received greater attention regarding influence of precision farming on their efficient use. Let us consider a few examples. Dobermann et al. (2004) have listed several instances pertaining to cereal production in Great Plains of USA, where in, precision farming has improved fertilizer-N efficiency. Adoption of precision techniques actually improved recovery rates and reduced NO_3-N leaching from the system. Similarly, precision techniques have improved fertilizer-N efficiency of rice grown in India, China, and other Far eastern regions. The agronomic efficiency of fertilizer-N (AE_N) was generally higher by 28% due to precision techniques compared to conventional farming systems. This had direct impact on dynamics and productivity of rice in the region. Precision techniques have generally envisaged application of relatively lower levels of nutrients, especially N. Clearly, N supply to soil is reduced without affecting grain/forage yield. This results in better agronomic efficiency of fertilizers. In case of maize, precision techniques has allowed harvest of 10–11 t grain ha^{-1}, yet reducing fertilizer-N supply by 10–13 kg N ha^{-1}. In case of wheat, fertilizer-N reduction has ranged from 9 to 51 kg N ha^{-1} for grain yield levels of 2.5–7 t ha^{-1}. In Asian locations, supply of about 32–45 kg N ha^{-1} could be reduced to rice crop grown under precision farming (Dobermann et al., 2004). The AE_N improved by 2–14 kg grain kg^{-1} N due to precision techniques adopted on wheat. The AE_N under conventional farming systems was 35–53 kg grain kg^{-1} N supplied. Precision farming increased AE_N by 16–42 kg grain kg^{-1}N in case of maize and by 12–28 kg grain kg^{-1} N in case of rice. Clearly, N use efficiency is a key parameter that precision technique affects in a given field or agroecosystem. In addition, Dobermann et al. (2004) have stated that sensor-based N management systems often indicated better N use efficiency compared to other methods of N management in fields. The N savings ranged from 10–20% with marginal or no increases in cereal grain yield. The extent of N savings, N use efficiency and profitability of sensor-based N management depended much on grain/forage yield levels and quality of the product. Whatever be the intensity of cropping, sensor-based N management did affect N dynamics in the field. It proportionately affects that of entire agroecosystem when extrapolated.

Precision farming approaches reduce loss of nutrients *via* percolation, leaching and erosion. This helps in enhancing nutrient use efficiency. For example, in case of P and K, adoption of precision techniques has improved fertilizer use efficiency, because it reduces loss of these elements from the field. Phosphorus recovery rates could be higher under precision farming leading to better fertilizer-P efficiency. Although relatively small residual P held in the profile may also add to fertilizer-P efficiency.

Precision Farming and Soil Physico-chemical Properties

Precision farming techniques are utilized to correct quite a few maladies that are primarily related to physico-chemical traits of soils. For example, variable lime inputs channel using precision techniques removes soil pH variations. Addition of variable rates of FYM based on soil maps reduces variations in CEC and buffering capacity of soils. Precision techniques could also be used to overcome variations in soil texture, tilth, and aeration. It requires use of appropriate soil maps and computer guided ploughing.

Precision Farming, Nutrient Transformations, and Soil Microbial Activity

Precision farming has direct effect on fertilizer and organic manure distribution in the soil profile. Variable rate supply of fertilizers and FYM removes variations in nutrients and soil organic fraction. Therefore, nutrient transformations that ensue could be expected to be proportionately uniform. There is no doubt that soil organic matter (SOM) distribution and concentration has immediate effect on soil microbes, their population and activity. Precision techniques, since they impart uniformity in SOM and microbial process, we may forecast that nutrient transformations too would be proportionately uniform in the soil profile. Repeated adoption of variable rate techniques and accurate placement of fertilizers plus FYM may induce stratification of microbial activity. Actually, microbial load and activity may just follow suit the nutrient stratification trends. Physico-chemical transformations that affect nutrient availability to roots may become more uniform in the profile. There is every possibility that microbes adapted to large variations in physico-chemical traits, nutrients and SOM get progressively less accentuated. Diversity of microbial species and successions may also be influenced by repeated adoption of precision techniques. Accurate placement and timing of FYM and inorganic nutrients may also restrict microbes, SOM and nutrient transformations to a particular zone in the soil profile. For example, relatively smaller quantity of nutrient loss/percolation to subsurface under precision farming may not support large population of microbes in lower layers. Microbial population may actually get localized more in the root zone. No doubt, variations in rates of nutrient transformations get minimized in response to uniform distribution of nutrients and soil microbes. Some of these aspects need to be experimentally tested. After all, microbes are highly versatile. Reports dealing with effect of precision farming on microbial flora suggest that bacterial and fungal diversity as well as population were not affected in the upper horizon of soil. The microbial community structure changed only in the later stages of crop. Microbial community seemed to be influenced more by root exudates and rhizosphere effect of plants kept under precision farming. The soil enzyme activity was influenced more by the seasonal changes and soil fertility differences that occur due adoption of precision techniques (Schloter et al., 2003). We have to accrue knowledge about influence of precision techniques on N mineralization and immobilization rate during a crop cycle. The influence of precision farming on microbial community relevant to nitrification and nitrification process is worth studying in detail. It has direct bearing on N cycle in the fields. Over all, a large body of knowledge about influence of precision techniques on SOM, microbial flora and their activity, nutrient transformation rates and nutrient availability to roots is yet to be accrued. Much has to be ascertained through careful experimentation. For example, bio-fertilizer application using Global Positioning Systems (GPS) guided traction may help us in maintaining uniformly optimum or high levels of N-fixers in the soil profile. We may also envisage variable rates of bio-fertilizers using their activity as criterion that is quantity of N-fixed per unit soil. Precision techniques could be used to distribute microbial inoculants containing biological nitrogen fixers. It may influence population, activity and contribution of such symbiotic microbes in soil.

Precision Farming and Nutrients in the Crop Phase

Precision farming may influence nutrients held in both soil and crop phase of an agrocosystem. Actually, precision techniques induce relatively more uniform crop growth and yield formation. It is a consequence of variable rate nutrient and water supply. Precision farming actually creates uniformity in nutrient and moisture availability to crop roots. Nutrients recovered and accumulated at various stages from seedling to maturity could be proportionately uniform. Nutrient distribution within the crop also gets more precise. Root to shoot ratio becomes uniform within the entire field. Harvest index too becomes more uniform across the field or cropping zone. It means extent of biomass available for recycling and that to be used as forage becomes more uniform in a field. *In situ* recycling may provide greater uniformity to SOM distribution. Since nutrients and FYM supplied are accurate, loss of SOC *via* soil respiration would be more even across the entire field. At the same time, C sequestration rates through root biomass could also be uniform and entirely dependent on quantity of residue recycling. Nutrients removed *via* grains or forage is uniform across the field or even agroecosystem. Of course, it is based on nutrient concentration in the aboveground biomass.

3.2 PRECISION FARMING INFLUENCES NUTRIENT DYNAMICS IN AGROECOSYSTEMS

3.2.1 Precision Farming and Dynamics of Major Nutrients

Maine et al. (2005) have stated that during past 23 decades many of the farming zones in the entire world perceived an enhanced use of fertilizer-based nutrient inputs, especially major nutrients– Nitrogen, Phosphorus, and Potassium. There was also a perceptible increase in irrigation within farming zones. High input farms needed extra supply of nutrients and water in addition to other inputs. Efficient utilization of fertilizer became a necessity, if farms had to be profitable. Reduction of ground water contamination meant accurate placement and utilization of fertilizer. Precise application and use of fertilizers was important. Hence, site-specific management of nutrients using GPS, Geographic Information System (GIS) and variable rate input was evaluated in many regions of the world.

Precision farming has both direct and indirect influence on various aspects of N dynamics in a crop field. If extrapolated, some of them like N supply or N loss may have far reaching effects on entire agroecosystem. Large-scale modification of aspects like N supply or removal from soil to crop phase or recycling influences variety of other ecosystematic functions, in addition to cropping pattern and productivity. Let us consider some of these aspects in greater detail in the following paragraphs.

Let us first consider N supply to fields that support either medium or high yield goals. Among nutrients, generally N is required in quantities more than others. Further, large scale deficiency and wide variations in its availability to crop roots is encountered in most agricultural belts. Fertilizer-N and/or organic manure supply is the most common method to correct such problems of N dearth. Farmers adopt many techniques to assess soil N fertility variations and to arrive at most appropriate fertilizer-N dosage. The fertilizer-N dosage is fixed considering, soil nutrient dynamics, crop response, and economic advantages. Nitrogen is major nutrient element that interacts extensively with soil phase. Such biological, physical, and chemical interactions affect both soil properties and N availability to crops. Therefore, predictions regarding crop's response to fertilizer-N inputs should consider various transformations that affect dynamics of N availability and absorption by crop roots. In case of precision farming, N inputs are guided by soil N maps and computer models. It also considers wide range of N transformations that occur in the soil profile. Such computer models consider extensive data that depicts aspects of N dynamics like different forms of soil N, N mineralization rates, N uptake rates, soil N retention, subsoil N accumulation, N seepage, percolation and emission rates, pre-planting soil organic N and inorganic N. A recent report by Delgado (2011) states that in USA, .a new GIS approach helps farmers with soil database. Information on soil N distribution could be downloaded and used as a map, to fix variable rate of fertilizer-N. Such GIS-NLEAP based techniques add accuracy to predictions about N dynamics. It also helps in soil N management during crop production.

Reduction in Nitrogen Supply

Dobermann et al. (2004) state that precision farming affects fertilizer-N supply to a maize field. The fertilizer-N inputs could be reduced by 9–12 kg N ha^{-1}. Yet, it keeps harvest levels at near optimum (10.5 t ha^{-1}). Similarly, for wheat grown in Great Plains of USA, N supply could be reduced by 9–32 kg N ha^{-1} without affecting grain yield (6–7 t ha^{-1}). In the Asian rice belt, precision farming could immensely influence N dynamics by reducing its input into field. Nitrogen impinged into rice belt could be reduced by 32–45 kg N ha^{-1} (Table 1). We should note that, N supply is key to overall N dynamics in any agroecosystem and this step is clearly influenced by adoption of precision techniques. The extent of reduction of N supply into cropping zones is of course important. It may have immediate effects on N turnover and recycling; N recovered into aboveground portions of the crop and ratio of N to other nutrients. It may ultimately affect crop biomass produced.

According to Babcock and Paustch (1998), shifting from recommendations that stipulate uniform application of fertilizers to site-specific variable rate inputs seems necessary due to various reasons. In addition to yield increase and input savings, variable rate technology (VRT) seems to be preferred, because it corrects nutrient dynamics. It avoids misapplication of N and corrects environmental influences of over loading of fields with fertilizer-N. It reduces wastage of fertilizer-N and avoids deterioration of aquifers and ground water. Simulations based on large number of field trials in experimental stations and farmer's fields suggests that single rate N or uniform application oversupplies maize fields with N. Adoption of VRT supplies quantities of N as required by crop. Therefore, VRT results in profits attributable to savings of fertilizer-N. Placement of fertilizer-N in exact.

TABLE 1 Reduction in nitrogen fertilizer supplied to cereal fields due to adoption of precision farming techniques some examples.

Crop, Yield (t ha^{-1}) **Location**	**Nitrogen Fertilizer Input (kg N ha^{-1}) Farmer's Practice**		**Reduction Variable-rate N**	**Reference of N input**
Wheat, 3.4 Oklahoma, USA	65	43	22	Arnall et al. 2010
Wheat 4.5–5.5 BioBio region, Chile	215	160	55	Claret et al. (2011)

TABLE 1 *(Continued)*

Crop, Yield (t ha^{-1}) Location	Nitrogen Fertilizer Input (kg N ha^{-1}) Farmer's Practice		Reduction Variable-rate N	Reference of N input
Wheat, 8.5	258	250	8	Mayer-Aurich et al.,
Potsdam, Germany 2007				
Wheat, 4.5–5.6	150	108	42	Dwivedi et al., 2011
Lohtaki, Haryana, India				
Wheat 7.4	174	155	19	Dobermann et al., 2004
United Kingdom				
Wheat 9.4	240	189	51	
Netherlands				
Corn, 9–10	235	195–220	10–50	Kitchen et al., 2009
Missouri, USA				
Corn, 7–8	185	160	25	Scharf et al., 2009a
Missouri, USA				
Corn 8.0	195	120	75	Miao et al., 2006
Nebraska, USA				
Rice 7.5	130	87	43	Dobermann et al., 2004
Philippines				
Rice 5.0	142	110	32	
Tamilnadu, India				
Rice 6.0	171	126	45	

TABLE 1 *(Continued)*

Crop, Yield (t ha^{-1}) Location	Nitrogen Fertilizer Input (kg N ha^{-1}) Farmer's Practice		Reduction Variable-rate N	Reference of N input
China				
Rice 9–10	197	174	23	Buresh et al., 2006
Vietnam				
Cotton, 0.75	55	49	6	Yu et al., 2001
Texas, USA				
Pearl millet, 3.1	125	114	11	Dwivedi et al., 2011
Lohtaki, Haryana, India				

Note: Farmer's Practice refers to quantity of inorganic fertilizer-N and organic N supplied and set of conventional or traditional agronomic procedures related to nutrient management. variable rate N refers to N supplied using variable applicator under precision techniques. Reduction refers to amount of fertilizer-N saved by using VRT compared to farmer's practices.

Reports from Argentina state that since fertilizer supply itself is marginal, VRT is not common. Quantities and close to roots is crucial. This aspect is achieved better under VRT than conventional methods.

Kitchen et al. (1994) reported that in maize growing regions of Missouri, USA, VRT generally provided better crop response compared to conventional approaches. Yield maps from previous seasons, grid sampling, soil analysis, and yield goals were used to decide fertilizer-N supply through VRT. Yield mapping enormously improved fertilizer-N efficiency and reduced loss of N to lower horizons or to atmosphere *via* emissions. Crop response and N efficiency were superior under VRT compared to fertilizer-N supplied based on "best years" or "best area". Therefore, N supplied to crops under precision farming was often perceptibly lower than in conventional systems.

Adoption of Site-specific Nutrient Management (SSNM) could easily reduce fertilizer-N inputs into the fields in several locations within the Corn Belt of USA, For example, Snyder (1996) have reported that corn grown under precision farming requires 3–13% less fertilizer-N to produce grain yield similar to farmer's practices. Miao et al. (2006) have shown that economically optimum rates of N (EONR) could be reduced by 69–75 kg ha^{-1}, through SSNM and split application of N to a corn crop that yield 8–8.5 t grains ha^{-1}. Ruffo et al. (2006) have also found that variable rate N inputs improve grain harvests. In fact, VRT can reduce fertilizer-N supply to a certain extent, without affecting grain SSNM, essentially provides uniform distribution of nutrients like N, P or K. It marginally reduces fertilizer-based nutrient supply into the cropping zone (Plate 1 and 2). SSNM reduces loss of N *via* natural factors. This is attributable to split and variable techniques that do not allow undue accumulation of nutrients in the subsurface. Ultimately, SSNM helps in reducing nutrient loss from the ecosystem. Of course, it imparts uniformity to crop growth and grain productivity. The SSNM reduces discrepancy in nutrient ratios through prompt changes to nutrient supply at each spot. It avoids nutrient imbalances by replenishing exact quantities of nutrients (Plate 1 and 2).

Scharf et al. (2009a) state that often, optimal fertilizer-N rates vary widely from year to year or field to field in a geographic location. One of the aims of precision farming is to avoid excess N supply to fields. In fact, crop sensors are often used under site-specific methods, so that N supply could be as accurate as possible based on season and specific field. Long term (5 year) trials in the "Corn Belt of USA" covering over 90 fields has demonstrated that sensor guided N and Producer's normal N rates both resulted in similar grain/forage yield. However, sensor-guided N supply saved on an average 22.6 kg N ha^{-1}. The fertilizer-N supply stipulated using crop sensors depends on soil moisture status and rainfall pattern during a year. For example, in a wet year fertilizer N recommended by sensors increased by 14 kg N ha^{-1} and it resulted in 8.5 bu ac^{-1}. Overall, long term assessments suggest that crop sensors could be useful component of SSNM methods. Crop sensors allow us to economize and reduce fertilizer-N inputs into the ecosystem.

In case of rice grown in South and Southeast Asia, the usual targeted agronomic N efficiency is 23 kg grain kg fertilizer-N applied to a crop that yields between 6.8 and 8.2 t ha^{-1}. According to Buresh et al. (2006), this is easily achievable by applying fertilizer-N at 70–131 kg N ha^{-1}. In the Red River valley of Vietnam, benefit due to site-specific N management was 0.7t grains ha^{-1}over farmer's practices. The requirement of fertilizer-N was reduced by 20–35 kg N ha^{-1}. In Philippines, rice farmers benefited by an increase in agronomic efficiency of fertilizer N. As a consequence, reduction in fertilizer-N use was 20–24 kg N ha^{-1} (Table 1). In India, reports suggest that an improvement in fertilizer-N efficiency resulted in reduction of fertilizer-N requirement ranging from 2 to 24 kg N ha^{-1}.

Nitrogen Efficiency Gains under Precision Farming

Nitrogen efficiency is an important aspect of N dynamics in a field/farm or even an enlarged agroecosystem. Nitrogen efficiency has immediate impact on extent of N inputs required to achieve a particular grain/forage yield targeted by farmers. Improved fertilizer-N efficiency reduces accumulation of N in the soil profile. Therefore, it avoids residual N that could otherwise be involved in biological and chemical transformations or be lost to the environment. Nitrogen use efficiency has two aspects. First one relates to N absorption from soil profile and other to biomass/grain production per unit N absorbed by crops. Nitrogen use efficiency also affects N recycling within a field or a farm. It has immense economic impact when extrapolated to agroecosystem or a crop belt. Fertilizer-N requirement could be proportionately curtailed, if N is recycled efficiently. According to Sudduth et al. (2010), one way to increase fertilizer-N efficiency is to adopt precision farming methods. Site-specific methods help us in distributing accurate amounts of fertilizer-N at each spot. Sub-field areas are easily marked and accurate quantity of N is supplied based on soil N status and grain/forage yield aimed. A relatively new approach is to apply fertilizer-N at various stages of crop growth (i.e. split dosages) using crop reflectance sensors mounted on to variable applicators. This procedure increases N available to crop roots and enhances fertilizer-N recovery rapidly at various stages. It leads to greater fertilizer-N efficiency. Standardized charts that depict correlation between crop reflectance pattern and plant N status are available. Split doses actually helps in matching N requirements of the crop at various stages through N supply. Together, crop sensors and split applications of fertilizer-N allows us to supply N at different rates. It also leads to uniform availability of N to roots at appropriate concentrations. It avoids both, excess or deficiency of N in the root zone of crops. According to Sudduth et al. (2010), farmers should be conversant with different types of crop sensors (GreenSeeker, CropCircle, CropSpec) and differences with regard to wavelength (visible or near infra red), orientation of sensors and calibrations in order to arrive at plant N status accurately.

PLATE 1 A Fertilizer spinner and Differential Global Positioning System (DGPS) fitted vehicle.

Note: The DGPS vehicles help in VRT nutrient application. They allow us to dispense exact quantities of fertilizer-based nutrients at each spot in the field.

Source: Mr. David Nelson, Nelson Farms Inc, Fort Dodge, Iowa, Iowa, USA

PLATE 2 A complex assembly of variable rate N, P, and K applicator ready to go into a corn field in the Corn Belt of Iowa, at Nelson Farms near Fort Dodge, Iowa.

Note: The VRT applicator has a fertilizer tank for each major nutrient, several connecting tubes, and injectors that coincide with rows. Quantity of fertilizer dispensed at each spot is directed by the computer-based decision support system. This allows farmers to apply variable amount of N, P, and K as required at each spot. Farmers derive a level of uniformity with regard to soil nutrient status. Generally, VRT decreases fertilizer-N requirement.

Source: Mr. David Nelson, Nelson Farms Inc, Fort Dodge, Iowa, USA

Computer based decision support systems used during VRT has a say in the amount of N distributed. Usually, computer models that accurately forecast crop's response to N inputs are selected. Data supplied to computer-based decision support systems are crucial. Fertilizer-N needs could get either over or under estimated based on data supplied. We may realize that even small errors, upon extrapolation to large cropping zones can cause significant alterations in N dynamics. During early stages of adoption of precision farming in the Great Plains region of USA, researchers tried to evaluate benefits of computer-based fertilizer recommendations and VRT, using simulations/crop models. Hypothetical conditions envisaged were applied to 10 different fields, with varying soil characteristics. Crop response to fertilizer-N was assumed to be initially linear that later plateaus at 100 kg N ha^{-1}. Forcella (1993) found that N supply using VRT was most effective, if soil fertility variations were conspicuous. The type of fertilizer-N used was indeed an important factor determining both N dynamics and profitability of precision farming procedures. Inexpensive fertilizers were cost effective. The VRT also reduced misapplication cost that occurs in conventional systems.

Agronomic efficiency of N (AE_N) is actually a cumulative effect of various natural phenomena and agronomic procedures adopted during crop production. Generally, AE_N improved due to precision techniques. Reports by Dobermann et al. (2004) have suggested that AE_N improves by 2–45 kg grain ha^{-1} N, depending on crop species and location.

Field evaluations in Missouri and Nebraska in USA have clearly shown that supply of fertilizer-N using active light reflectance (crop sensors) and variable rate N applicators, immensely affects N dynamics in the corn field. Recent reports suggest that adoption of such site-specific management method reduces fertilizer N usage by 1050 kg N ha^{-1}. Clearly, accurate timing and placement of N as splits using crop sensor readings helps in efficient N use by the crop. It avoids undue accumulation of N in the profile that could otherwise be vulnerable to loss *via* percolation, seepage or emission (Kitchen et al., 2010; Roberts et al., 2009; Table 1). Nitrogen removed *via* crop residue and/or grain would be proportionately affected.

Godwin et al. (1999) evaluated two methods of VRT. First one involved "on-the-go" site-specific N input based on sensor readings. Second method was based on historic data and maps prepared using soil analysis. Yield maps pertained to a 3 year wheat-barley rotation followed in the fields in Iowa. Fields were treated after marking them into series of management strips. As expected, strips supplied with highest N rates using VRT method provided best harvest. These results were commensurate with soil fertility status and yield goals envisaged. The VRT based on site-specific variations of soil N performed better than one based on historic data. However, interaction between precipitation pattern, weather in general and VRT-N inputs needed due attention.

In Iowa, corn farmers practice precision farming techniques rather routinely. However, they consistently search for methods that allow reduction of fertilizer-N supply, without affecting crop growth and grain yield markedly. Reduction of fertilizer-N supply has immense effect on soil N accumulation. Undue accumulations induce proportionately greater loss of N *via* percolation, erosion and emissions. High soil N is also of detriment to ground water quality. Kyveryga et al. (2010b) have reported a study where in the aimed at reduction of fertilizer-N supply by one-third of originally applied through variable rate N applicators. They examined the effect of reducing fertilizer-N supply by 56 kg N ha^{-1} to corn field. Simulations and actual field trials indicated that yield reductions were significant, if normal moisture supply was not altered. Economically significant yield reductions occurred when rainfall during the season was 300 mm. Kyveryga et al. (2010b) have actually described a set of 22 field trials that evaluated and quantified the extent of fertilizer-N reduction possible using site-specific methods compared with normal rates. Results indicate large variations in yield reduction that occur due to reduced fertilizer-N supply. Obviously, within field variations were significant. In fact, the extent of yield reduction due to reduced fertilizer-N supply was greater, if it was a rainy year or rainfall was > 300 mm. It means interaction between fertilizer-N and irrigation is important. Similarly, organic matter content of the soil was also a factor that interacted with fertilizer-N and affected crop response to N inputs.

Runge and Hons (1998) too have pointed out that variation in nutrient elements in the soil is not the foremost factor affecting corn yield. Instead, interaction of various other factors such as, N x irrigation also affects crop growth and yield. Often, advantage of variable N inputs is affected rainfall pattern in a given location.

According to Khakural et al. (1996), corn yields were affected by extent of soil erosion. Nutrient loss from the field was itself influenced by landscape, its topographical features and extent of erosion. Often, corn yield was greater at foot slope positions and side slopes. Therefore, soil slope gradients and N distribution needs careful evaluation prior to fertilizer-N inputs *via* variable rate N applicators. Such details could be carefully included in the models used in decision making. Clearly soil factors other than just nutrient distribution in the profile and availability do affect crop growth and productivity.

Reports by Jiyun and Cheng (2011) suggest enhanced N recovery from soil and better fertilizer-N efficiency are the major causes for increased grain/forage yield in the maize/wheat growing regions of China. The fertilizer-N use efficiency improved from 30.3% to 40.0% due to adoption of SSNM. Clearly, it induces definite changes in N dynamics within soil profile. Recycling of biomass and N too gets affected. Most importantly, soil N distribution and fertility in general gets more uniform.

Precision Farming Reduces Nitrogen Loss from Fields/Ecosystems

According to Fiez et al. (1994), in addition to economic benefits, VRT-N based on yield potential avoids both over and under supply of N to wheat fields. Field investigation in different topographic locations (slope or depression) and fields with different depths of soil showed that, VRT-N avoids N loss, because almost every bit of fertilizer-N applied is utilized by wheat crop. There is no misapplied N or over dose of N that commonly occurs with single rate N supply techniques. Farmers essentially need prior information about rooting pattern and wheat grain yield potential in that location, N requirement per unit grain yield, residual-N or inherent soil N and N recovery rates of the crop. No doubt, accurate recommendations and variable rate N inputs are crucial to obtain better N dynamics in the field. It is said that under VRT-N, wheat grain production is slightly low but this is offset by reduced N supply. Accumulation of N in soil profile and recycling are generally controlled better, if VRT-N is adopted.

Burton et al. (1999) studied the impact of variable rate N inputs and rainfall pattern on corn production in Tennessee. They used EPIC crop model to simulate weather patterns and calibrate fertilizer inputs. Large 40 ha plots with differences in soil fertility were studied for corn productivity with uniform N rates and VRT. Farmers opted to use VRT when rainfall patterns were more suitable for maize production. Adoption of VRT provided better crop response to N input and profits even when rainfall was relatively low. Rainfall affected net N loss from the soil profile, carryover of N (residual N) and N recovery from soil to crop phase of the ecosystems. The VRT lessened loss of N from the soil profile to drains. Since, N did not accumulate in the soil profile to any great extent, its loss *via* emission, seepage or percolation was minimal.

Usually, N that accumulates in soil or that held as residual N is vulnerable to loss *via* erosion or emissions. Precision farming actually stipulates application of accurate quantities of N that match the needs of crop. Hence, N accumulation or residual N in the profile is minimal. As a consequence, N loss through soil erosion and runoff are minimal. Since NO_3 accumulation is minimal under precision farming, its loss too is low compared to conventional systems. Let us consider an example from farming zones of Chilean highlands. According to Claret et al. (2011) the residual N accumulation in soil profile was proportionate to fertilizer-N supplied to wheat fields. Quantity of fertilizer-N supply explained about 98% of the variation in residual N. Farmers thought that soil N fertility would build up a residual N accumulated in the subsurface and surface layers. A small fraction of residual N was immobilized but sizeable amount was prone to leaching and emission loss (Table 2). Lower rates of fertilizer-N application resulted in reduction of N leaching. Therefore, N-loading of groundwater was avoidable. Further, it was found that maximum leaching occurred immediately after first irrigation (35–43%). It seems plausible because during early seedling stage fertilizer-N requirements are low and a large fraction that goes unused becomes vulnerable to leaching. Hence, Claret et al. (2011) suggest that precision techniques that envisage application of smaller quantity of N and in split dosages based on soil fertility maps should be practiced. Split application improved fertilizer-N use efficiency. Reports indicate that precision farming, firstly improves N dynamics during wheat cultivation by matching N need with demand at various stages of the crop. Therefore, it leaves very little residual N that could be leached or volatilized. Further, precision techniques allow reduction in fertilizer-N supply to an extent of 55 kg N ha^{-1} compared to 215 kg N ha^{-1} applied under farmer's practice (Claret et al., 2011).

Precision Techniques, Landscape, and Soil Types

Bruulsema et al. (1996) have reported that in maize fields of Minnesota, a three-way interaction between topography, soil N indices and VRT affected crop response. Soil N indices or landscape characteristics alone did not affect soil N recovery or crop response. Management strips located in low-lying areas gave better correlation between VRT-N and crop response. Strong yield response to N in low-lying areas was attributed to accumulation of nutrients, organic matter and water. Interestingly, crop response to VRT-N inputs correlated better with soil N indices, especially total N, compared to N supply based on economically optimum rates. Therefore, N dynamics is affected by the location and VRT.

TABLE 2 Precision farming reduces loss of nitrogen from wheat fields in chilean highlands.

Farming Practice	Fertilizer N applied to crop	Nvolatilized from soil	Nextracted by crop	Nleached from soil	Nresidual in soil
			kg ha–[1]		
Farmer's Practice	215	6	122	8	165
Precision Farming	160	5	128	6	107
SPAD Chlorophyll Meter Readings	91	4	123	7	44

Source: Claret et al. (2011)
Note: Average mineralized-N was 64 kg N ha–[1] that added to variation in soil N distribution. SPAD refers to N supply based on sensor readings.

Roberts have evaluated integration of active sensors directed on a crop (maize), grown on management zone. Such an integrated system is supposedly more robust during decision making and variable rate N inputs into maize fields. It definitely improves fertilizer-N distribution in soil and its efficiency. Field trials in Nebraska during 2007–2008 have shown that, on both fine textured and eroded sandy loams, integrating soil management zones, and sensor-based N inputs resulted in N savings. Since these are initial findings, it is believed that further calibration may offer greater quantity of N savings. We should note that such saving in N input has direct impact on N loaded into soil profile, its distribution, extent of percolation and emission loss. Finally, it affects the extent of N traced in the ground water. The amount of N removed into crop can be regulated better by integrated sensor-based management zone systems. Overall, any improvement in assessing N needs of a crop, obtaining better distribution in soil and efficient use by plants ultimately means more favorable impact on N dynamics in maize belts.

Precision Techniques, Fertilizer-N supply, and SOM status

Graham et al. (2010) have argued that in the humid tropics, N management during corn production is often imprecise and inefficient. Nitrogen management techniques rely on expected economic benefits, yield goals, soil test, and crop N sensor readings. However, according to them, factors like weather parameters, soil organic carbon, mineralization trends, hydrology, and agronomic procedures may all affect soil N availability to maize roots. Therefore, simulation model that considers N dynamics in greater detail has to be used in the decision support system. It has been suggested that models that simulate N transformations, soil N, and water transport in the profile, corn crop growth, N uptake trends at different stages are more relevant. They offer greater precision regarding supply of N to corn fields. Clearly, N supply based on prior knowledge about SOM and carbon status measured using near-infrared (NIR) reflectance spectroscopy has its advantages in regulating N dynamics in the corn field.

Fertilizer-based N supply definitely induces higher crop production. Yet, it also enhances CO_2 and N_2O emissions. In some cases, N accumulation leads to eutrophication of water bodies near farm and/or ground water contamination. Scharf et al. (2009b) believe that precision farming could be adopted, since it envisages supply of exact quantities of fertilizer-N as needed by the crop, based on yield goals. It also ensures a certain degree of protection against undue accumulation of N in the soil profile or ground water. Further, precision farming removes within field variability for soil N and other nutrients. Greater detail on effect of precision N management and its influence on N and C dynamics in the crop field/ecosystem are highly useful. At this juncture, knowledge about influence of precision farming on C:N ratio in the field seems useful. Farms that confine precision techniques to fertilizer-N inputs, may still experience variations in C:N ratio encountered by crops. Precise adjustments in C:N ratio could help us in regulating N mineralization/ immobilization patterns and N availability to crop roots. Although not common, it is useful to trace variations in SOC and apply SOM accordingly to fields using variable rate applicators. Maintaining uniformity with regard to C:N ratios and correcting variations in both and soil N and SOC may be a good idea if economically feasible.

3.2.2 Phosphorus and Potassium Dynamics under Precision Farming

One of the basic steps in precision farming is to develop a soil map of a field depicting variations in P and K. Farmers can also manipulate P and K supply if they possess suitable and sufficient historical data regarding distribution of these two elements in the field. Soil P and K are well distributed both in surface and subsurface layers of soil. Therefore, accuracy of P and K data has often depended on inclusion of quantity of these two nutrients held in subsurface layers of soil. Let us consider an example from Northern Great Plains where wheat or maize is grown after demarcation of management zones or strips. Hammond and Mulla (1988) developed a map based on soil samples derived at 100–400 ft intervals in the management strips. They could categorize fields into low, medium and high for P and K contents. Conventional or uniform P and K resulted in over fertilization of 45% area and caused under supply in 8% of 50 ha field. Whereas, VRT supplied quantities of P and K exactly required by the crop. Clearly, soil mapping, categorization and VRT do affect P and K dynamics. Residual, P and K was minimal.

During recent years, precision farming techniques have been extensively used to assess variations in soil P, its availability, absorption pattern by crops and consequent effects on grain/forage yield (Christy et al., 2010; Mallarino et al., 1998). Unlike fertilizer-N, in most situations, fertilizer-P inputs are channeled as a basal doze in one stretch. Further, fertilizer-P added is often not efficiently absorbed by crops. Since, only 22–25% fertilizer-P is used by the first crop, it leaves a certain amount of P in soil as residual P. The influence of residual P on nutrient dynamics in soil, crop productivity and its economic value should be considered accurately. Let us discuss a few examples where precision faming approaches have been used successfully to supply fertilizer-P to different crops. Mallarino et al. (1998) have shown that strip farming methods common to Corn Belt of USA, could be used to grow corn/soybean intercrops and apply precision farming techniques. They used 20 ha strips to evaluate corn and succeeding soybean crop in Iowa, USA. Maps depicting P availability in soil were prepared using Bray-1. They evaluated a set of three treatments namely non-fertilized control, fixed P (conventional approach) and variable rates of P applied to soil, based on previous data or soil chemical analysis. Fixed rate was 52 kg P ha^{-1}. Variable rate of granulated monoammonium phosphate was incorporated into soil using a truck equipped with DGPS receiver and a flow controller. Variable rate of P applied ranged from 38 to 58 kg P ha^{-1} depending on soil data. Grain yield was monitored using combines equipped with real-time DGPS receivers. Flow rate sensors (e.g. AgLeader-2000) were used to prepare yield maps. Following data explains effect of variable rate inputs of fertilizer-P.

Crop		**Corn**			**Soybean**	
Fertilizer-P Input ▶	Control	Fixed	Variable	Control	Fixed	Variable
Grain Yield (kg ha^{-1})	9039	9182	9321	4017	4118	4080
SE±	86.9	70.1				

Source: Mallarino et al. (1998)
Note: Fixed = conventional fertilizer-P supply

In case of corn, variable rate P supply improved grain yield in addition to imparting uniformity to soil P fertility. Profitability of precision farming could also be observed in terms of reduction in P supply and improved fertilizer-P efficiency. Average fertilizer-P used reduced by 2–11 kg P ha^{-1}, if variable rate was adopted compared to conventional application of fertilizer-P at seeding. Clearly, precision farming affects both P dynamics in the field and crop (corn/soybean) productivity. In the Corn Belt, especially in USA, farmers find that the effect of VRT-P or VRT-K is highly dependent on extent of low-P or low-K zones found in the field. The benefit from VRT, in terms of better P and K dynamics and crop response occur when a field possesses 20–80% low-P and/or low-K zone. Of course, factors like grid size, number of soil samples collected, analyzed, and fertilizer schedules also affect benefits derived from VRT on maize production. Computer models used to generate fertilizer recommendation and yield goal also influence benefits from VRT. Field validations in Kansas have indicated that "on-the-go" soil P (Mehlich's P) analysis and recommendations based on appropriate decision support systems are possible (Christy et al., 2010).

Report from Maize growing regions of Ohio in USA suggests that intensity of grid sampling and estimation of soil P are crucial factors that affect our understanding about the extent of soil P variations. Grid sampling seems to have far fetching effects on P supply, its depletion and recycling in the field. Actually, sampling distance affects the soil P estimates. For example, Beuerlein and Schmidt (1993) have reported that soil P estimates vary between 27 to 38 kg P ha^{-1} if grid sampling is adopted. However, soil P estimates would vary enormously between 11 to 70 kg P ha^{-1}, If only one sample is studied $hectare^{-1}$. Larger grid sizes mute the variations pertaining to soil P if any. The intensity of grid sampling affects various other decisions pertaining to fertilizer-P inputs, amount of P held as residual pool, subsurface P, recovery of P by the crop and yield. Economic evaluations suggest that grid sampling and P estimations cost about 4–4.5 US$ ha^{-1} more than conventional methods that cost 50 US$ ha^{-1}. Wells and Dollarhide (1998) found that soil P and K was affected by the quantity of fertilizer-P and K added through variable rate techniques. Fertilizer-P and K inputs using "on-the-go" devises improved soil P and K test but decreased soil pH.

Griffin et al. (2000) studied the influence of VRT on P dynamics in the rice fields of Arkansas. They used EPIC crop model to recommend fertilizers to rice-soybean rotation. Three different treatments of conventional uniform P inputs were compared with a VRT based on soil analysis. The VRT-P supply affected both soil P dynamics and improved crop response. The VRT was suitable only when soil P fertility was highly variable and soil types too varied within a field. The VRT was not desirable if soil types were uniform within a field or management strip. Instead, uniform rate or single rate P supply was congenial, if soil type and soil P fertility was relatively uniform.

Farmers in North America or elsewhere in other continents usually manage P and K dynamics carefully, in order to slowly build up their reserves in the soil profile (Lowenberg-DeBoer and Reetz, 2002). In some regions with intensive farming trends, farmers may aim at rapid improvement of P and K fertility status in soil. Mostly, farmers tend to supply P and K fertilizers based on State Agency Recommendations or Best Management Practices. We should note that during first year only a fraction of P and K is recovered by the crop. A sizeable fraction stays in the soil profile and is known as residual P or K. This fraction of residual P and K gets depleted as cropping proceeds, if proper replenishment plans are absent. However, repeated supply of P and K fertilizers into the field improves their (P and K) content and availability to crops that are grown in succession. During this process of P and K accumulation in the soil profile, if proper care is not taken, then within field variations get expressed more conspicuously through the years. As such, crops grown may deplete P and K at different rates causing variations in P and K availability in a field. In addition, the natural process like chemical fixation, transformations, erosion, percolation, seepage, and loss *via* drainage may all be uneven and variable spatially and temporally. These factors again induce marked variations in P and K fertility. It is useful to adopt precision farming techniques and supply P and K at variable rates. This leads to a semblance of uniformity in soil P and K fertility. We can therefore forecast that soil P and K fertility build up in field would be uniform in case of management strips, fields or large expanses that are treated with precision farming. There are procedures that allow rapid buildup of P and K in soil. There are others, which enhance P and K content slowly in soil, through many years of farming. Slow build of P is common in low input or subsistence farming regions. Slow build up occurs when quantity of residual nutrients left in the field after a crop is relatively small. During this period of P and K build up, adoption of precision techniques helps in retaining uniformity in P and K distribution.

Computer-aided simulation and reports from practical farming situations suggest that precision farming techniques could be utilized to regulate soil P status during sugar cane production in Brazil. Actually, precision techniques allow us to avoid both over and under application of fertilizer-P to the Dystrochrepts. These soils are generally highly variable with regard to soil P availability (Sparovek and Schnug, 2001). Over application of fertilizer-P is common in 73% of the area where conventional methods are adopted. Following is an example that depicts sugarcane yield gain due to precision farming:

Treatment	**Cane Yield**	**Relative Yield**
	t ha–^{1}	**%**
No Fertilizer-P	57	71
Precision-P supply	75	93
Conventional-P supply	74	92

Source: Sparovek and Schnug, 2001

Precision techniques have direct impact on K dynamics of Oxisols that support soybean production in the Cerrados of Brazil. Soybean grain yield varies widely between 2 and 6 t grain ha^{-1}. It is attributable to soil fertility variations, especially P, and K. Traditional farmer's methods that envisage fixed rates of fertilizer-K supply seem to underestimate crop's demand for K. Whereas, variable rate K supply ensures uniformity in soil K availability. Most importantly, precision techniques help in balancing soil K removed *via* soybean grain with that applied based on yield goals. Hence, in a long run, precision techniques avoid soil K depletion and reduction crop productivity.

Precision techniques, when applied to large tracts of Brazilian coffee growing regions, promises to affect nutrient dynamics, especially that of P and K positively, in terms of soil and agro-environmental parameters. Reports have already shown that coffee farmers would reap almost similar levels of coffee seeds but with a reduced supply of P and K fertilizers to soil. For example fertilizer-P supplied under uniform rates (traditional farming) is 417 kg P ha^{-1} plus 387 kg K ha^{-1}. This gets reduced to 320 kg P ha^{-1} and 334 kg K ha^{-1}, if VRT are adopted (Molin et al., 2010; Table 3). Reduction in chemical fertilizer impinged to soil leads to smaller quantity of residual P and K in subsurface layers. Nutrients vulnerable to loss *via* percolation and seepage would be proportionately lesser.

Reports from South African farming zones indicate that precision farming techniques are mostly utilized to supply fertilizers and lime as accurately as possible. However, there are large farms that adopt precision farming to regulate P and K dynamics. Maine et al. (2007) evaluated over 180 experimental strips supporting maize, for response to fertilizer-P that was either supplied as single rate or applied using VRT. Results suggested that VRT was more efficient and it related to better P recovery by crop. Variable rate inputs gave higher profits than single rates.

Grain Research and Development Agency of Australia has conducted series of field trials in the Australian wheat belt using precision farming techniques to study fertilizer-P requirements (GRDC, 2010). Reports indicate that farmers save about 4–6 kg P ha^{-1} by adopting precision farming. When extrapolated, P supply into wheat fields or large farms or cropping belt will get proportionately reduced. For example, farmer owning a 1000 ha farm lessens fertilizer-P supply by almost 4,000–6,000 kg into the region. In addition to P dynamics, interactions of P with N and water could also be affected due to variable supply of nutrients. Further, if fertilizer-P is efficiently used then residual P levels could decrease. Actually, P inputs get reduced plus it is absorbed well by the wheat crop (Table 3).

TABLE 3 Reductions in Phosphorus Fertilizer application due to precision farming approaches.

Location/Soil Type/Crop	Reduction in Fertilizer-P Supply kg ha^{-1}	Reference
Southern Australia/	46	GRDC, 2010
Red Earth, Wheat		
Ames Iowa, USA/	211	Mallarino et al. 1998
Haplustolls, Maize	614	,,
Sao Paulo state, Brazil		
Hapludoxisols, Coffee	98	Molin et al. 2010

Note: Reduction in fertilizer-P refers to difference of P input between farmer's traditional practice and precision technique.

Pearl millet-based cropping systems common to Western Gangetic plains were examined for effects of SSNM, SAR (State Agency Recommendation), and Farmers Practice (FP) on soil nutrient status and nutrient dynamics. Report by Dwivedi et al. (2011) indicate that organic C and N were affected least and accumulations, if any, were marginal at the end of the cropping system. Whereas, soil P status post-wheat or post-mustard enhanced by 5 kg ha^{-1} in fields under SSNM compared with FP or SAR. Wheat or mustard fields kept under SSNM showed a buildup of available K but those under FP or SAR got depleted of K. The residual K status of SSNM, SAR, and FP treated plots were as follows:

	Residual K status (kg ha–^{1})	
Treatment	**after Mustard**	**after Wheat**
SSNM	+24	+12
SAR	–13	–17
FP	–18	–25

Source: Dwivedi et al. 2011

Note: Positive sign refers to accumulation and negative to depletion of available K in soil at the end of crop sequence.

It is clear that soil P and K dynamics are affected to a certain extent due to cropping. Adoption of SSNM affected P and K dynamics differently compared to SAR or FP. SSNM resulted in optimum crop yield plus allowed a small quantity of P and K to accumulate. Fields under SAR and FP resulted in depletion of P and K. We should also note that generally, SSNM stipulates addition of major nutrients to correct deficiencies as exactly as possible, plus micronutrients in appropriate proportions. This leads to maintenance of nutrient balance in the cropping ecosystem. Nutrient ratios in the soil are held optimum under SSNM. In case of both SAR and FP, only major nutrients were supplied and at rates not exactly matching the deficiencies and needs of the crop in question. Hence, nutrient dynamics was affected adversely leading to incorrect ratios and lack of nutrient balance.

According to reports by IPNI Asia program situated in Gurgaon, India, adoption of SSNM to maintain optimum K dynamics in the southern Indian plains has been necessitated due to soil K dearth and wide spread variations in crop response to fertilizer-K. For example, Satyanarayana et al. (2011) state that high rate of K removal; uneven depletion due to cropping and uneven loss of K due to natural processes has lead to variation in soil K availability in fields. Factors like cropping systems; climatic variations, and improper fertilizer-K input schedules have also induced soil K variations. During past decades, farmers generally resorted to uniform application of K decided by state agencies. Such recommendations were dependent on soil test for K; yield goals and a tendency to replenish K to balance the removal. These procedures did not remove variations in soil K availability. Currently, farmers growing field crops have been utilizing SSNM procedures. Procedures like soil tests for K that includes measurement of available K, exchangeable K, and preparation of soil K maps are routine. Computer-based decision support systems that are endowed with models like "*Nutrient Expert*" or "*Nutrient Manager*" are being used to decide fertilizer-K schedules. Fertilizer-K distribution is done using GPS guided farm vehicles. This seems to establish uniformity with regard to soil K. It affects K dynamics positively and maintains soil K balance. Long-term effect of precision techniques on soil K dynamics is worth investigating through field trials.

3.2.3 Precision Farming influences Dynamics of Secondary and Micro-nutrients

Sulfur nutrition needs greater attention in case of oilseeds because these crop species demand higher quantities of S for grain formation and fat accumulation. As such, sulfur distribution in soils could vary enormously. It is distribution, both in the surface and subsurface layers of soil need to be assessed. We ought to realize that many of the deep rooted crops, more so oilseed crops extract substantial amounts of S trapped in subsoil layers. Hence, in case of fields where variable rate S technology is adopted it is customary to feed data pertaining to both surface and subsurface layer. Haneklaus et al. (2002) have suggested that often it is ratio of N and S in soil that affects crop yield. In other words, variations in ratio of available N:S is important. They report that soils found in German plains vary for both S and major nutrient N. Sulfur status varied between 40 to 70 kg S ha^{-1}. Spatial variability of soil moisture and certain physico-chemical traits also affects sulfur distribution and its absorption by crops. Oilseeds show greater response and profitability for variable rate S supply. Further, they have suggested that SOM content, N mineralization rates should be considered along with variations found for S availability in soil. We may note that, in soil, greater portion of S utilized by crop is originally found in organic form. Actually, variations in both available N and S fluxes need due attention. Further, keeping continuous information on soil S variation through time seems most useful.

There are several reports that suggest that sulfur deficiency and variations in distribution and availability affect crop production in the Eastern European cropping zones. For example, Vanek et al. (2008) suggest that spatial variability for S is felt across cropping zones in Czeck Republic, yet it is not as rampant as that traced for soil N. The soil S content both in surface and subsurface layers need to be assessed and data should be fed to decision support system, prior to variable rate application. The soil S variations have been felt more in areas devoid of industrial emissions.

Field trials on Ustochrepts found in North Indian plains have clearly shown that adoption of SSNM does affect S dynamics in the field. In a pearl millet–mustard rotation, post-mustard available S in soil was maintained at initial levels in fields kept under SSNM. Whereas, fields under SAR or FP showed an average decline of 6.7 kg S ha^{-1} (Dwivedi et al., 2011). In a pearl millet-wheat rotation, post-wheat residual S was 2.2–8.0 kg S ha^{-1}. The subsurface S was an important factor for deep rooted crops like mustard. A sizeable fraction of S was contributed by subsurface S. Hence, computer models preferred during variable rate S supply should be amply corrected to consider subsurface S.

There are indeed innumerable reports about methods of sampling and analyzing plant or soil samples for Ca and Mg, in order to map variations in their concentrations within a field. We can match availability of soil Ca and Mg with plant tissue tests. Quite often, plant tests for Ca and Mg coincide with soil pH. However, low Ca and Mg does not always connote that soil pH would be low. Also, high Ca and Mg do not mean higher soil pH (Franzen and Swenson, 1995). Soil maps depicting Ca and Mg distribution could be prepared and fed to decision support system in order to apply variable rate of dolomite (or gypsum) based on exact requirements. In general, GIS based soil maps for secondary and micronutrients are available. They depict variations in nutrient distribution plus other factors that are deemed to influence availability and absorption of secondary and micronutrients. Such maps, if prepared periodically allows us to monitor dynamics of secondary and micronutrient in each field or even on a wider scale in an ecoregion. For example, soil maps that depict Zn distribution and factors that may affect its availability are available for most states in India (Bali et al., 2010). Planners could easily obtain a broad idea about variations in Zn status and distribute Zn-fertilizers accordingly to overcome variations. Such data is available for many regions of the world and can be easily retrieved *via* internet or networked computers. For example, secondary and micronutrient status, and recommendations for various counties of states like Iowa (Sawyer et al., 2008), Nebraska (Wortmann, et al., 2008) is available. Similarly, such data is available for different agroecological regions of India (Singh, 2001; Wortmann, 2008). Similar data for other cropping zones of the world are also available.

3.3 PRECISION FARMING: DYNAMICS OF SOIL ORGANIC CARBON

Remote sensing is a valuable technology to study SOM distribution. Digital imagery and computer software required calibrating soil colors and organic matter is often minimal. Digitized version of soil maps could be used to study SOM content and mineralization rates in time and space. The cost of generating a SOM map is relatively small and data could be used effectively for computer-aided decision support system. However, high intensity grid sampling and analysis are costly (Schepers, 1996; Sudduth et al., 1990).

The influence of precision farming is easily felt in management zones that are generally highly productive and possess optimum or high organic C. According to Moshia et al. (2010) variable rate N inputs based on soil N estimates may not provide expected results if soils are sandy, low in organic matter (FYM) and occur in management zone, that are generally deemed low productivity zones. They suggest application of organic manure to supply most or part of N as organic manure in order to improve soil quality and response to variable rate N technology. This step actually corrects C:N ratio in precision fields. Further, based on series of trials in Colorado, it was suggested that application of organic manure alone using variable rate techniques does not suffice. A portion of N should be applied as inorganic fertilizer to prime the crop. Actually, they opt for a strategy wherein a part of N is supplied as organic manure and rest as inorganic fertilizer-N. In a way, farmers would then be regulating both C and N dynamics in the field more precisely.

3.4 PRECISION FARMING AND BIO-FERTILIZERS

Soil fertility variation within a field is quite common. Firstly, it connotes variations in distribution of soil inorganic nutrients and organic matter. It also includes variations in soil microbial population and their activity. Incidentally, nutrient transformation rates are immensely regulated by soil microbial flora. In the present context, we are concerned more with symbiotic N-fixers like *Bradyrhizobium* and asymbitoic N-fixers like *Azospirillum, Azotobacter* and so on. The species diversity of N-fixers and their activity in soil is highly variable in most fields. It means extent of N derived from atmospheric N-fixation within a field varies proportionately. Perhaps we could apply precision techniques and correct the variations in N derived *via* soil microbes. It involves soil maps that depict N fixation trends. Then, data on N fixation pattern could be supplied to variable rate applicators through a decision support system that considers soil microbial population, their activity and N derived from atmosphere. Computers should also consider factors that influence over all process of biological N fixation. These operations are costly, time consuming, and cumbersome right now. Economic feasibility of such an effort needs careful consideration. Soil maps depicting variations in soil microbial population and their activity can be prepared. It may be costly yet possible. In fact, there are techniques allow us to selectively map the microbes related to bio-fertilizers (N-fixers or P solubilizers) in terms of population, diversity, and activity. We can also prepare soil maps that exclusively depict variations in symbiotic and asymbiotic N-fixers in a field. Variable rate techniques could then be adopted based on microbial activity maps. Farmers may actually decide extent of N derived from bio-fertilizers. In due course, repeated adoption of precision technique may stabilize a field with bio-fertilizers and extent of N derived from atmospheric fixation. Precision techniques could be used to supply fields with uniform levels of bio-fertilizers. Over all, we may strive to derive precise quantities of N from atmospheric N fixation. The process of estimating variations in microbes and their activity could be costly and cumbersome at times. Perhaps there is a strong need to devise techniques that allow "on-the-go" estimation of soil microbial density and accordingly direct variable rate applicators to supply microbial inoculums. This is easily said than done. Biofertilzer bags come with data about microbial (N fixers) density and N fixation activity. Hence, variable rates of liquid inoculums could be dispensed appropriately using soil maps and decision support systems. Perhaps basing bio-fertilizer inoculums release on soil respiratory activity (tetra zolium test) and N fixation activity (acetylene reduction) could become feasible in due course of time. Knowledge about localization or stratification of N fixation activity due to adoption of precision farming is useful. Soil organic matter distribution and rooting pattern may actually affect distribution of microbes contained in a bio-fertilizer preparation. Precision farming using management blocks or strips and manual application of biofertilizer is feasible at present. Farmers can achieve a semblance of uniformity of bio-fertilizer distribution in a management strip or block.

Reports from European farming zone suggest that precision farming may not affect soil microbial community structure and their activity in the top layers of soil. The enzyme activity in field kept under precision techniques was affected by the nutrient input rates, season, root exudation, and rhizosphere effect (Schloter et al., 2003). Clearly, establishment of microbial inoculants, its perpetuation and activity are influenced by precision farming through its effect on various soil physico-chemical effects and crop species.

3.5 PRECISION FARMING AND ITS IMPACT ON SOIL PROPERTIES (TEXTURE, SOIL REACTION, ELECTRICAL CONDUCTIVITY, SOIL MOISTURE, TOTAL CARBON ETC.)

Investigations on variability of soil nutrient status, moisture, crop growth, and yield are important while developing a site-specific management strategy. At the same time, attention to soil factors like drainage, runoff, erosion, and pH are as important as knowledge about soil nutrient status and availability to roots (Karlen et al., 1998). Let us consider an example. Paddy is major cereal crop in many of the Southeast nations. In the Fareast, Korean farming zones too support large patches of paddy cultivation. Fertilizer supply is generally high ranging from 250 to 320 kg N, P, and K ha^{-1}. Crop productivity is proportionately high because of intensive farming trends. Soils used for paddy cultivation vary enormously for various characteristics like nutrient content and availability to crop, soil moisture, pH, electrical conductivity and so on. Chung et al. (2008) point out that during precision farming, obtaining data about variations in soil properties is important. Conventional sampling and chemical analysis in laboratory may be difficult to be applied for soil properties. Instead sensors seem more promising in identifying the variations in soil properties. In fact, in some areas, farmers have been exposed to sensors that use visible and NIR wavelength to estimate soil properties in a given management zone. Calibrations made on soil samples derived from 65 cm cores have clearly shown that reflectance measurements can help us in estimating sand, silt and clay fractions, pH, EC, P_2O_5, Na, Mg, Ca, and total C. The basic idea is to adopt *in situ* close range sensors that can supply useful data to farmers prior to using variable rate applicators.

Precision Farming to Correct Variations in Soil Reaction (pH)

Soils vary for traits that directly relate to fertility like nutrients and moisture. They are also highly heterogeneous for traits like soil pH, electrical conductivity, and redox potential. We should note that aspects like nutrient availability, nutrient transformations, occurrence of different chemical forms of nutrient elements and rate of nutrient acquisition by crops are all influenced, throughout the season by soil pH. Hence, it is worth studying and identifying the within field variations of soil pH and correcting them, using precision techniques. Soil pH is supposedly one of the most variable traits encountered by farmers. Soil pH is also a trait that has immediate implications to fertilizer efficiency, nutrient dynamics in general, and therefore ultimately grain productivity. Within the realm of agricultural cropping, soil acidity is commonly indicated by soil pH. It is a measure of hydronium ion activity in the soil solution. The primary question here is how does soil pH get altered to different extents in certain locations and not so in others? Rather, what are the factors that contribute to soil pH variations that need to be corrected in the first instance? As an alternative, soil amendments could be supplied to match the soil pH variations in a field. This may lead to uniform soil pH in a field. Indirectly, it ensures normal and uniform availability of soil nutrients to crops planted in a field. Incidentally, crop roots are highly susceptible to variations in soil pH. Root activity relevant to nutrient absorption too gets affected as soil pH varies from one location to other within a single field.

Commonly known reasons for variations in soil pH or creation of soil acidity are:

(a) Removal of bases by harvested crops, leaching of bases, acid residues left in soil profile due to fertilizer inputs, and inherent nature of soil parent material (Bongiovanni and Lowenberg-DeBoer, 2001).

Liming is an important time tested procedure to correct soil pH. Lime is usually supplied just prior to planting. It is supply is based on current soil pH, soil volume to be corrected, and extent to which soil pH needs to be enhanced. Liming may not be a necessity in all cropping zones. In most cropping zones soil pH may actually be highly congenial to the crop species planted. Therefore, it is influence on nutrient acquisition rates may go unnoticed, even if any. Incidentally, liming is an important procedure in several cropping belts. For example, areas in Great Plains of North America, Cerrados, and Pampas in South America, large tracts in European plains, South Asian Red Alfisol/lateritic regions and Chinese farming zones may all need periodic soil pH correction. Most cropping zones require some amount of lime as amendment, if soil pH recedes below 5.0. However, if soil pH rises 7.5, it may then make it too basic. Lime can inhibit crop growth if used rampantly and can add to variations in grain productivity. Therefore, during variable rate inputs of lime, quantity to be dispensed should be carefully and accurately estimated. During recent times, site-specific or precision techniques have been adopted to supply lime that corrects soil pH and brings about a degree of uniformity to crop productivity.

Bongiovanni and Lowenberg-DeBoer (2001) evaluate the effects of variable rate lime inputs to fields meant to support corn production. The acidic haplustalfs found in Indiana are affected by soil acidity and they require supply of lime to correct it. Soil pH is a trait highly variable even within a field. Hence, proper judgment and variable rate supply are required, if uniform soil pH is to be achieved. Totally, they adopted four different strategies to correct soil pH variations within a field. First one is SSM based on agronomic recommendations provided by the agricultural agency of Indiana. Lime requirement for desired pH was calculated based on lime index and yield optimization. Here, economic aspects are not given importance. It seems currently, in the Corn Belt of USA, SSM based on agronomic considerations are most popular. The second strategy involved SSM with due consideration to economic aspects like efficiency of lime inputs and profitability. It is similar to SSM based on agronomic consideration for the most part but uses economic parameters like marginal value of the product, maximization of net profits, and profitability on large scale use. The third strategy adopted by Bongiovanni and Lowenberg-DeBoer (2001) involved consideration of large body of information about the field, its location, cropping history, and lime input technologies, generally used by corn growers in Indiana. The other aspects considered were cost of variable rate equipments, cost of lime application and value of the product. The three strategies were compared with whole farm management (WFM) techniques that are in vogue in Indiana since decades. The WFM system involves soil sampling, chemical analysis of composite samples and single application of lime to entire farm/field based on recommendation. The baseline results indicate that SSM with agronomic consideration provides best results in terms of soil pH uniformity, soil fertility and crop productivity. Following is an example that depicts the extent of liming necessary in different locations in a corn growing state like Indiana in USA and expected profits due to adoption of precision farming.

TABLE 4 Influence of Site-specific nutrient management (Precision techniques) on lime requirement and profitability.

Location	WFM	SSM–Agronomic	SSM—Economic	Information
Southwestern Indiana				
Lime Input t hectare^{-1}	0.49	1.13	1.53	1.30
Profitability US$ hectare^{-1}	385	391	395	388
Davis Farm, Indiana				
Lime Input t hectare^{-1}	1.68	1.92	2.58	1.49
Profitability US$ hectare^{-1}	421	434	438	436
Lynn Coop, Indiana				
Lime Input t hectare^{-1}	2.27	3.53	2.38	0.96
Profitability US$ hectare^{-1}	410	434	441	435

Source: Bongiovanni and Lowenberg-DeBoer, 2001

Note: WFM = Whole Farm Management (Conventional); SSM = Site-specific Management; Information = Lime applied based on Information about Profitability of SSNM. Profits due to adoption of Precision farming range from 13 to 25 US$ ha^{-1}.

Overall, we should note that obtaining uniform soil pH in the field is a requirement. It may have direct influence on several physico-chemical properties of soil, nutrient availability and absorption rates. The SSM strategies bring about greater uniformity to soil fertility, hence their influence on nutrient dynamics in the soil and crop productivity could be different from those noted for WFM techniques. Let us consider an example from Southeast USA (Florida), where in Ultisols are acidic or variable in pH and need regular supply of gypsum once in 23 years to achieve uniform soil reaction. Heiniger and Meijer (2000) examined over 110 fields in the coastal zones for pH. They picked samples using 1.0 ha rectangular grids and soils were variable with regard to pH. Application of gypsum according to state recommendations or uniform rate did not remove variation in soil pH entirely. Gypsum was either over or under supplied in many locations. Lime requirements actually increased under uniform rate compared to grid sampling and variable rate supply. In the Coastal Plains of Georgia, regular use of VRT was useful in correcting soil pH accurately and all across the field. The VRT-pH also reduced gypsum requirements, hence cost of production decreased proportionately. We should note that pH affects dynamics of many soil nutrients. In particular, aspects like nutrient availability to roots, fixation into soil matrix and transformations are all influenced by soil pH. This is in addition to effect of soil pH on root activity and nutrient recovery rates.

3.7 PRECISION IRRIGATION AND ITS IMPACT ON NUTRIENT DYNAMICS

Within this section, influence of precision irrigation on soil nutrients and crop productivity has been discussed through examples derived from North America and other regions. Let us consider an interesting example from Canadian Prairies. This region supports wheat and canola production on undulated landscape with knolls and depressions. Dynamics of both water and N seems most crucial to crop yield. The glacial soils of Saskatchewan in Canada show large variations in soil moisture retention, SOM, and N distribution. Here, nutrient and moisture dynamics is immensely influenced by topography. Precision techniques may be helpful in overcoming certain variations in soil fertility and moisture availability. Under natural conditions, water is lost rapidly from knolls and a drier condition develops, thus limiting the net crop response to other factors like variable supply of fertilizer-N and FYM. The lower slopes and depressions that accumulate water, organic matter and eroded nutrients exhibit significant variations in soil fertility, and crop growth. Clearly, management zones could be easily demarcated into knolls, mid slopes and depressions or troughs. Field scale investigations by Walley et al. (1998) suggested that on the knolls, precision techniques might correct N variations, yet the crop productivity may not show pronounced response because water becomes limiting. Precision farming techniques seem to be most suitable in the depressions, where soils are rich but variable with regard to nutrients, organic matter, and moisture. Supply of water and fertilizers were based on aerial photographs and variable rate applicators. Evaluations indicated that variable rate fertilizer-N supply should be considered in areas where responses are greatest that is in the depression. Water and organic matter are not limiting factors in the depressions. Seed yield of canola consistently increased due to adoption of precision techniques but correction of both N and water dearth seems important. Further, we should note that precision techniques are worthwhile in areas with crop species that respond to variable rates of fertilizer and moisture supply. For example, in the present case, extent of response of wheat to variable rate nutrient supply was not sufficient enough to support it adoption. Overall, it is soil nutrient availability, presence of water in non-limiting levels and a crop species that responds to precision farming techniques that clinches the issue, whether to invest on such techniques or not. Long-term effect of precision techniques on nutrient dynamics at the knoll and depressions is worth watching. We may forecast that farmers would benefit by reducing undue accumulation of nutrients and organic matter in the depressions. Since exact quantities of fertilizers are applied on the knolls loss due to erosion too would be reduced on a long-term basis (Figure 1).

Field trials in several locations within in the Great Plains area suggest that irrigation directly affects nutrient loading into the drains and channels. Adopting precision agriculture could regulate nutrient loss from wheat fields. Several studies have compared precision irrigation with uniform rate of water supply as envisaged by the State Agricultural Agencies. In the present case, "Best Management Practices" was usually considered as equivalent to uniform rate technology and compared with VRT. Drainage water, seepage, and percolations were tested for nutrient loading and net loss of fertilizer-N and other nutrients. Since, precision techniques supply nutrients in quantities exact enough to meet the yield goals, excess nutrients are generally absent and are not lost *via* drainage. In fact, experiences suggest that in most farms practicing both VRT nutrients and VRT for irrigation, net loss of nutrient from field is very small. Clearly, precision irrigation and VRT Nutrients affects soil nutrient dynamics immensely. The consequent effects on crop growth and nutrient recycling could be easily calculated using appropriate models.

Indeed, there are several reports on usefulness of precision irrigation to rice crop production and nutrient dynamics in the Asian wetland regions. Regulation of water and nutrients is a major preoccupation during lowland rice production. In California, rice crop encounters soil fertility variations that are primarily attributable to nutrient loaded into irrigation water that moves into rice fields. It is said laser leveling and strict water management is still insufficient to overcome the variations in nutrient availability (Simmonds et al., 2010). Water salinity too induces a certain degree of variation in soil fertility. It has been suggested that in order to achieve proper redistribution of nutrients, it is preferable to obtain maps that depict nutrient distribution. Then, use appropriate decision support systems and variable rate applicators to channel major nutrients N, P, and K. In addition, it is preferable to monitor and regulate nutrient concentrations in the irrigation water.

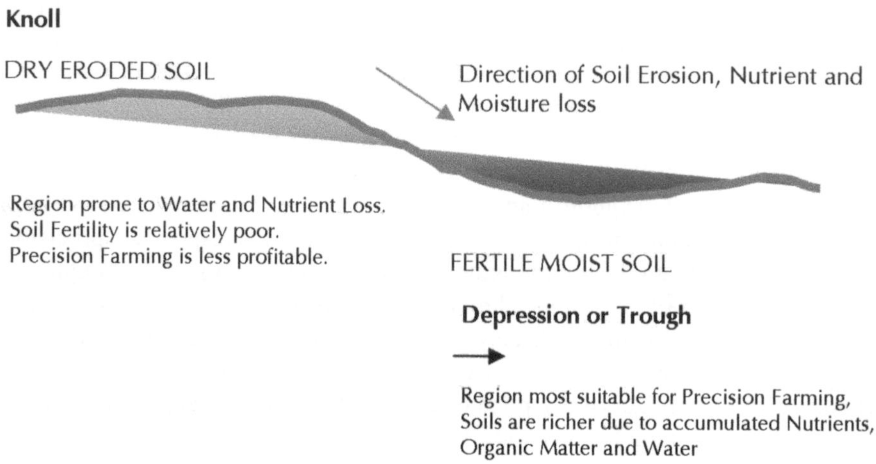

FIGURE 1 Topography, Water and Nutrient Interactions that ensue in the Canadian Prairies affect adoption and Profitability of Precision Farming.

Let us consider an example from cropping zones of Egypt. El-Nahry et al. (2010) compared response of maize to traditional and precision farming techniques on a sandy soil in the Ismailia region of Egypt. Precision techniques, as usual, included grid sampling, analysis, and VRT for both fertilizer and water supply. At least three management zones were identified to supply nutrients and water at different rates. Fertilizer inputs were made according to soil tests and yield goals. Irrigation was done based on variations in soil water retention capacity and depletion rates by the crop. It was suggested that adopting precision techniques for both nutrients and water was important. It reduced loss of nutrients and avoided deterioration of soil and aquifers in the farming zones. Precision techniques were invariably needed to maintain uniform moisture availability to roots. Actually, on sandy soil, factors like runoff, seepage, percolation emissions, and uneven absorption by crop roots may all contribute to soil fertility variations.

Precision Farming, Weeds, and Crop Productivity

During practical farming, weeds deplete soil nutrients along with crop plants. Weeds are among the important factors that deplete soil fertility unevenly and induce variability. The extent of nutrient depreciation by weeds varies with weed density, species, its stage, foliage, flowering pattern, and seed production. Actually, weeds divert nutrients meant for main crop in the field. The extent of nutrients diverted away from main crop varies, based on many factors that influence competition between weeds and crop. Such competition too causes variations in soil fertility, photosynthetic rate, and biomass production by the crop. Generally, weed distribution in a cropped field or that kept under unplanted-fallow is highly variable. Herbicide application has to be directed based on weed intensity. Hence, during recent years, VRT has been applied advantageously to apply herbicides in appropriate quantity based on maps that depict weed infestation. Over application of herbicides can result in environmental hazards and under application might cause uneven weed removal and crop yield loss (Bradley and Khosla, 2003). It is interesting to note that map-based weed removal resulted in accurate herbicide sprays and reduced its consumption by 6–69% depending on fields. Herbicide costs were significantly reduced. Sensors that rapidly identify weeds are available. They utilize radiation reflected from leaf surface to identify weeds.

Precision Farming and Long-term Nutrient Dynamics

Longterm effects of precision farming are as important as immediate gains that farmers accrue in terms of reduction in fertilizer inputs, uniformity in soil nutrient and organic matter distribution, regulation of crop growth, and improvement in grain yield. In a long run, precision farming does affect various parameters of agro-environment like soil nutrient dynamics (nutrient loss to lower horizons, deterioration of ground water, nutrient loss *via* emissions, residual nutrients), mineralization/immobilization trends, soil water resources, SOM and microbial distribution in the soil profile, crop residue recycling trends, and grain harvests. Many of these long-term effects of precision farming on agro-environment are specific to geographic location, cropping pattern, and economic situations. Also, we must realize that a great number of advantages that may occur due to precision techniques, in comparison to farmer's traditional systems are actually attributable to uniformity in soil fertility that it creates and reduction in nutrient/water supply to the fields. Accurate data on some of these aspects need to be accumulated *via* experimentation. There are not many long-term studies. Computer aided simulations and forecasts could be useful guides about long-term effects. In general, since precision technique stipulates application of accurate quantity of fertilizer and irrigation, residual nutrients are minimal.

In a long run, influence of precision technique on residual nutrients will be almost negligible, if fertilizer inputs match removal by crop. Precision farming stipulates accurate placement and timing of fertilizer-based nutrient at a particular depth in the soil profile. This practice may induce nutrient stratification. Precision farming involves accurate timing of split applications of fertilizer-N and irrigation events. Firstly, it creates uniformity in the entire field with regard to nutrient and water availability. Next, percolation of nutrients to lower horizons could become minimal if nutrient scavenging by crop root system is efficient. Precision farming, in fact avoids ground water contamination and delays soil deterioration. Again, since, residual nutrients are minimal, loss of soil N due to NO_2, N_2O or N_2 emissions, and NH_3 volatilization is markedly curbed. Similarly, loss of P *via* chemical fixation and runoff could be reduced due to precision techniques. Precision farming may induce stability with regard to crop residue recycling and soil organic matter status. Decision support systems may direct accurate inputs of FYM based on SOM maps. Therefore, soil chemical and microbiological aspects are influenced. The rates of nutrient transformation may stabilize at a particular level as cropping pattern, nutrient inputs, removal and recycling trends get standardized and microbial flora gets stabilized. Nutrient transformation rates also depend on fertilizer inputs and crop yield goals. However, long-term adoption of variable rates induces uniformity with regard to nutrient transformation and turnover rates in a field or management zone. This needs to be confirmed through experimentation. Longterm adoption of precision farming may after all reduce costs on fertilizer inputs, irrigation, and offer greater profits to farmers. However, we ought to realize that economic advantages from precision techniques may also taper off in a long run. The relative gains over farmer's techniques may not be as great as it is during initial years of adoption of precision techniques. Over all, precision technique offers a great opportunity to regulate soil fertility and nutrient dynamics in any agroecosystem. It also provides us with a method to delay or minimize or even halt totally the deterioration of soil and other resources of a cropping zone.

Precision Farming and Nutrients in the Agroecosystems: Extrapolations

The impact of precision farming on nutrient dynamics and ecological functions depends on scale at which these advanced satellite-guided techniques are adopted. Viglizzo et al. (2004) argue that intensive cropping and application of precision techniques in smaller zones like a field or plot may induce changes specific to small area. However, scaling up precision techniques to agroecosystems at large may bring about ecological changes of greater significance, of course along with greater profits that may accrue to farmers. This aspect needs to be simulated and tested by whatever means possible. Prior knowledge based on evaluation of authentic data about influence of precision techniques on entire agroeosystem is indeed very useful. Let us consider some interesting examples and hypothetical situations in the following paragraphs. These examples deal with fertilizer-N which is perhaps most important input in any agroecosystem.

Firstly, let us consider N dynamics in the wheat belt of Great Plains of North America. Farmers generally apply 180–240 kg N ha^{-1} and derive 4.0–4.5 t grain ha^{-1}. Reports from Oklahoma and other locations in wheat growing regions of Great Plains up to Canada in the North suggest that, a shift from farmer's practice to precision techniques reduces N supply into the agroecosystem by 30–55 kg N ha^{-1}. Clearly, N impinged into agroecosystem is lessened. If we extrapolate to entire wheat region of USA, reduction in consumption of fertilizer-N due to precision farming is 106 m t per season. It proportionately reduces soil N and fertilizer-N that is vulnerable to loss *via* emissions, surface flow, erosion, and seepage. Further, we may note that accurate placement of fertilizer-N at an appropriate depth in soil and in close proximity to wheat roots reduces extent of N emissions and volatilization. Total reduction in N loss *via* emissions reaches high proportions each season in the wheat belt of Great Plains of USA. Precision farming makes farmers apply exact quantities of fertilizer-N at each spot based on crop's requirements. As such, extraction of fertilizer-N applied becomes more efficient. It leaves very little scope for residual N to build and become vulnerable to loss through various natural processes or to weeds during fallow. Under Precision farming, crop residue recycling is more streamlined. The C sequestration trends and SOM build up or loss is controlled based on FYM prescriptions made by decision support systems. The C sequestration could be improved due to accurate placement of FYM. Precision techniques avoid undue and uneven accumulation of SOM. This helps in reducing loss of soil C *via* respiration. The ground nutrient dynamics of wheat belt to could be affected due to adoption of precision techniques. Rooting and nutrient depletion trends get more uniform in the wheat cropping zones. Overall, adoption of precision farming may induce greater uniformity to soil fertility and productivity of wheat belt. Plus, it reduces deleterious effects of rampant fertilization into the fields. This is in addition to economic gains that it may bestow to the farmers of this region.

The spread of precision technique in the intensively cropped "Corn Belt of USA" is impressive. It supposedly lessens use of fertilizer-N by 10–75 kg N ha^{1} depending on actual yield goals and location (Kitchen et al., 2009; Miao et al., 2006). The "Corn Belt" extends into 3.0 m ha and scale up effects on soil quality, agro-environment, and grain yield seems laudable. Fertilizer-N usage in the entire Corn Belt reduces by 150 m t annually (Table 1). In due course, precision techniques could spread over the entire corn belt of China. At 5% fertilizer-N reduction due to precision techniques it results in 23–40 m t fertilizer-N savings. Obviously, it avoids soil deterioration and restricts loss of fertilizer-N into atmosphere, aquifers and ground water. Generally, advantages from precision techniques are accentuated under high input, intensive cropping systems. However, there are good examples from regions with medium levels of fertilizer input and productivity. For example, fertilizer-N supply to Corn Belt in India is about 80 kg N ha^{-1} and productivity is proportionately lower at 2–3 t grain ha^{1} (Table 1). Farmers actually adopt several types of production practices. Hypothetically, precision techniques may offer a net 5–10% reduction in fertilizer inputs in comparison to other procedures. It is equivalent to 48 kg N ha^{-1} and when extrapolated to entire Corn Belt (7.0 m ha^{-1}) in India, it amounts to reduction of 28–56 m t of fertilizer-N. Precision techniques may offer a method to achieve greater uniformity in soil fertility and productivity.

Rice is a staple cereal in South and Southeast Asia. Here, rice cropping extends into 39 m ha. The grain and forage yield goals are relatively high. Therefore, it is grown intensively by using higher amounts of fertilizer, organic manures, and irrigation. For example, in South India, "Rice Agroecosyste" extends into 7.8 m ha. Alfisols, Coastal sands, and laterites that support rice production are all replenished consistently, season after season, with 80–120 kg N ha^{-1}. Farmers adopt a wide range of production packages like State Agency Recommendations, Maximum Yield technology, and Traditional Farmer' Techniques. Precision farming is the latest technique that is gaining in area. It promises to reduce fertilizer-N consumption. In the general course, about 10–12% fertilizer-N is lost *via* emissions and erosion, and 3–8% *via* NH_3 volatilization (see Krishna, 2003). Since fertilizer-N is applied accurately at each stage of the crop and placed closely to roots, it supposedly reduces loss of fertilizer-N, that otherwise occurs under usual farmer's practices. The net savings of fertilizer-N due to precision farming ranges from 23 to 45 kg N ha^{-1} depending on location and yield goal (Buresh et al., 2006; Dobermann et al., 2004). The scale-up effects of precision farming on the entire South Indian Rice belt should therefore be impressive. If extrapolated, it means we are reducing fertilizer impingement into rice agroecosystem by 265 m t yearly (Table 3). Repeated adoption of precision techniques can immensely thwart deterioration of soils, aquifers and agro-environment. Yet, it keeps total productivity of cropping zone at par with other production systems. Most importantly, farmers stand to gain economically by using up proportionately lower amounts of fertilizer-N.

Rice-wheat rotation practiced in Northern Indian plains extends to over 11 m ha. It is relatively intensively cropped zone that yields over 7 t grains ha^{-1} annually to farmers. Farmers adopting traditional techniques impinge their soils with 180–230 kg N ha^{-1} annually. Since fertilizer-N efficiency is 38% with rice and 43% with wheat, it leaves a sizeable fraction of fertilizer-N that could be vulnerable to loss *via* emissions, erosion, and seepage. Adoption of precision techniques may involve initial investments. Yet, it offers advantages in terms of reduction in fertilizer-N supply and delay in soil deterioration effects, if any. If extrapolated, at about 10% fertilizer-N efficiency due to precision techniques, fertilizer-N saved amounts to 200–280 m t annually in the entire Gangetic belt (Table 3). Precision technique also reduces loss of fertilizer-N. The economic advantages due to repeated adoption of precision techniques could be impressive.

Precision farming techniques may bring about significant changes in soil fertility status, nutrient dynamics, and productivity within the vast soybean producing zones found in the Cerrados of Brazil. At present, farmer's method consistently underestimates fertilizer requirements, especially P and K. Hence, soil productivity is low and variable. Nutrient recycling too seems low. Dearth for P and K is easily felt. The law of minimum takes effect and it limits soybean crop yield. In addition, it affects dynamics of other nutrients. Instead, if precision techniques are adopted, it firstly corrects soil fertility variations (P and K), regulates nutrient depletion, and replenishes soil P and K, in quantities that match their removal through grains. When extrapolated to vast soybean producing zones in Brazil, it is a significant impact on nutrient dynamics within the agroecosystem. Soybean productivity of this zone could improve perceptibly.

We should note that precision technique does not always mean high capital and initial costs. Certain variations may involve much less labor and machinery. Handheld instruments and manual spread of fertilizers using management blocks reduces investments. Practically, advantages from precision techniques could also be reaped in cropping zones supplied with low or medium levels of fertilizers and FYM. The net advantages may be proportionately low. For example, groundnut belt in Senegal, pearl millet in Northwest India, sorghum/legume intercrop in the Vertisol zones of South India are all good candidates for adoption of precision farming (see Krishna 2003, 2008, 2010). The fertilizers and chemicals impinged into these belts could be reduced conspicuously. Yet, keeping the grain/forage yield levels optimum. Let us consider the situation with sunflower in Peninsular India. It is cultivation has spread rapidly into Vertisol regions of Central and Southern Indian plains. In fact, a vastly stretched sunflower agroecosystem of 2.6 m ha has taken root in this area. Farmer's practices envisage application of 60 kg N and 30 kg P ha^{-1} (Table 1; Krishna, 2010). Adoption of precision techniques helps in removing the unevenness in soil fertility. It improves fertilizer efficiency because decision support systems allow accurate distribution of fertilizers at various stages of crop growth. It also considers the subsoil fertility, since sunflower is a deep rooted crop. The buildup of residual N may get minimized. The loss of fertilizer-N due to natural factors could be vastly reduced in the entire sunflower belt because under precision farming nutrients are garnered rapidly. Assuming a 10% advantage due to precision farming over farmer's traditional packages means it results in reduction of fertilizer-N supply to sunflower fields by 13.2 m t annually in the sunflower belt (Table 3.)

Application of fertilizer-P too could be reduced by 6–7 m t annually within this sunflower region. However, we may note that fertilizer-P is not entirely exploited in any cropping zone. It does leave a certain amount of residual P that could be utilized later. It is feasible to adopt both precision techniques involving elaborate GPS guided tractor driven seeders and fertilizer supply systems or hand-held sensors and manual distribution. Farmers in the rain-fed zone may adopt hand-held sensors or maps and those with affordable resources may opt for elaborate mechanization and decision support systems. Whatever is the scale of investment and sophistication in instrumentation, precision techniques promise to impart a certain degree of uniformity to soil fertility and sunflower productivity. It lessens fertilizer and chemical inputs into the agroecosystem. Precision techniques avoid nutrient accumulation and deterioration of aquifers in the sunflower belt. Long-term effects and gains in terms of soil fertility and grain production need to be examined.

Citrus is an important plantation crop of Florida. It spreads into 584,000 ha. The average fertilizer-N supply into citrus region is 200–270 kg N ha^{1} plus that derived from mineralization of FYM. Loss of fertilizer-N *via* volatilization is 10–40 % and that *via* erosion is 5–20% depending on location. Adoption of precision techniques may reduce fertilizer-N consumption to a certain degree. Adoption of variable rate techniques possibly imparts greater uniformity and accuracy with regard to fertilizer placement and supply levels. Fertilizer-N supply to Florida citrus is about 146 m t yearly. At an assumed 10 % advantage (approx. 25 kg N ha^{-1} less) due to precision techniques, citrus farmers may have to apply 14.6 m t fertilizer-N less than at present. Precision technique helps in reducing residual N. Hence, loss of fertilizer-N to lower horizons and ground water is reduced.

Molin et al. (2010) have found some interesting results regarding P and K dynamics within the coffee cultivating regions of Brazil. The reduction in P and K requirements attainable due to adoption of precision techniques is sizeable at 98 kg P ha^{-1} and 53 kg K ha^{-1}. When extrapolated to large coffee belt, precision techniques will definitely affect soil P and K dynamics per square. It reduces undue accumulation of P and K in subsurface, it avoids percolation seepage and run-off loss. It reduces the risk of contamination of drainage line with fertilizer-based nutrients. Most strikingly, it brings about uniformity of soil P and K status, therefore, we can expect uniform

TABLE 5 Impact of precision farming on nitrogen dynamics within an agroecosystem: extrapolations—some examples.

Agroecosystem	Area m ha^{-1}	Grain Yield t ha^{-1}	Fertilizer Farmer's Practice	Nitrogen Rate Reduction due to Precision Farming	Net Reduction of Fertilizer-N supply to Agro-ecosystem (m t)
			kg N ha^{-1}		
Corn Belt of USA	3.0	7 9	180 240	1075	150
Chinese Corn Belt	2.6	7 9	195	918*	2347
Corn in India	7.8	2 3	80	5	40
Winter Wheat in					
Great Plains of USA	37	45	180	1836*	106
Rice-Wheat Sequence	11	7 8	180240	1824*	200-280
in North India					
Rice in South India	7.8	1.7 4.2	60110	2345	265
Sunflower in South India	2.6	23	6080	68*	1621

Sources: Arnall et al. 2010, Buresh et al. 2006; Dobermann, 2004; Kitchen et al. 2009; Krishna, 2003; 2008; 2010; 2011; Tucker et al. 1995; Wang et al. 2006.

Note: Experimental data about fertilizer-N reduction due to adoption of precision techniques is absent in some cases. Values with * mark are assumptions that precision techniques may provide 10% advantage over traditional farmer's practice. However, there are situations where precision technique has no comparative advantage in terms of fertilizer-N needed or it could be higher than under farmer's practice.

crop growth and seed yield. Recycling of P and K would be nominal. Residual P and K would be manageable.

Overall, precision techniques seem to offer advantages in terms of soil fertility, fertilizer and water supply, and productivity in different cropping belts. It is impact on the entire agroecosystem and on a long run needs to be simulated, understood, and experienced in reality in due course.

KEYWORDS

- **Conventional system**
- **Differential global positioning system**
- **Farmyard manure**
- **Agroecosystem**
- **Near-Infrared**
- **Precision farming techniques**
- **Soil organic matters**

REFERENCES

Arnall, D., May, J., Butchee, K., and Taylor, R. *Evaluation of Sensor based Nitrogen application in Producers fields*. International Annual Meetings of American Society of Agronomy, Long Beach, California, USA, 107(5), pp. 1–2 (2010), Retrieved from http://a-c-s.confex.com/crops2010am /webprogram/Paper58587.html (January 4th, 2011).

Babcock, B. A. and Paustch, G. R. Moving from Uniform to Variable fertilizer rates on Iowa corn: Effects on rates and returns. *Journal of Agricultural and Resource Economics*, **23**, 385–400 (1998).

Bali, S. K., Kumar, R. Hundal, H. S., Singh, K., and Singh, B. GIS-aided Mapping of DTPA extractable Zinc and Soil characteristics in the State of Punjab. *Journal of Indian Society of Soil Science*, **58**, 189–199 (2010).

Beuerlein, J. and Schmidt, W. *Grid Sampling and Fertilization*. Ohio State University Experimental Station. Agronomy Technical Report No 9302, pp. 1–12 (1993).

Bongiovanni, R. and Lowenberg-DeBoer, J. Nitrogen management in Corn using Site-Specific crop response estimates from spatial regression model. *Proceedings of the 5th International Conference on Precision Agriculture*. Minneapolis, USA, pp 1–8 (2001).

Bradley, K. and Khosla, R. The role of Precision Agriculture in Cropping Systems. *Journal of Crop Production*, **9**, 361–381 (2003).

Bruulsema, T. W., Malzer, G. L., Davis, P. C., and Copeland, P. J. Spatial relationships of Soil Nitrogen with Corn yield response to applied Nitrogen. *Proceedings of 3rd International Conference on Precision Agriculture*. Minneapolis, USA, pp. 505–512 (1996).

Buresh, R. J., Pampolino, P. S., Tan, R., Rajendran, R., Gines, H. C., Son, T. T., and Ramanahan, S. *Reaching towards Optimal Productivity*. International Plant Nutrition Institute, Norcross, Georgia, USA, pp. 1–8 (2006), Retrieved from http://www.ipipotach.org/pt/eifc/2006/10/15.htm (June 30th, 2011)

Burton, E., Mahajanashetti, S. B., and Roberts, R. K. *Economic and Environmental benefits of Variable rate application of Nitrogen to Corn fields: Role of Variability and Weather*.

Proceedings of the American Agricultural Economics Association. Nashville, TN, USA, pp. 22–24 (1999).

Christy, C. D., Drummond, P., and Lund, E. *Precision Agriculture applications of an 'on-the-go' soil infra-red reflectance sensor*, pp. 1–12 (2010), Retrieved from http://www.veristech-com/pdf_files/ Optical_8thinticonf.pdf (June 12th, 2011).

Chung, S., Sudduth, K., Jung, Y., Hong, Y., and Jung, K. Estimation of Korean Paddy field soil properties suing Optical reflectance. In *Asabe Annual International Meeting Technical Papers*. American Society of Biological Engineers Annual International Meeting. Providence, Rhode Island. Paper No. 083682, pp. 1–3 (2008), Retrieved from http://asae.frymulti.com/abc.asp?JID=5&AID=25021&CID=prov2008&T=2 (December 15th, 2010).

Claret, M. M., Urrutia, R. P., Ortega, R. B., Stanely, B. S., and Valderrama, V. N. Quantifying Nitrate leaching in irrigated Wheat with different Nitrogen fertilization strategies in an Alfisol. *Chilean Journal of Agricultural Science*, **71**, 148–156 (2011).

Delgado, J. A. *A new GIS approach to assess Nitrogen Management across the USA*, pp. 1–8 (2011), Retrieved from http://www.icpaonline.org/finalpdf/abstract_176.pdf (June 21, 2011).

Dobermann, A., Blackmore, S., Cook, S. E., and Adamchuk, V. I. Precision Farming: Challenges and Future Directions. In *New Directions for a Diverse Planet*. Proceedings of the Fourth International Crop Science Congress, Brisbane, Australia, p. 19 (2004), Retrieved from www.cropscience.org.au (January 20th, 2011).

Dwivedi, B. S., Sing, D., Swarup, A., Yadav, R. L., Tiwari, K. N., Meena, M. C., and Yadav, K. S. On-Farm Evaluation of SSNM in Pearl millet-based Cropping systems on Alluvial soils. *Indian Journal of Fertilizer*, **7**, 20–28 (2011).

El-Nahry, A. H. Ali, R. R., and El Baroudy, A. A. An approach for Precision farming under Pivot irrigation systems using Remote Sensing and GIS techniques. *Agriculture Water Management*, **98**, 517–531 (2010).

Fiez, T. E., Miller, B. C., and Pan, W. L. Assessment of spatially variable nitrogen fertilizer management in winter wheat. *Journal of Production Agriculture*, **7**, 86–93 (1994).

Forcella, F. Value of Managing Within-field Variability. *Proceedings of the 1st Workshop on Precision Farming*. American society of Agronomy, Madison, WI, USA, pp. 125–132 (1993).

Franzen, D. W. and Swenson, L. J. Soil sampling for Precision farming. *Sugar beet Research and Extension Reports*, **26**, 129–134 (1995).

Godwin, R. J., James, I. T., Welsh, J. P., and Earl, R. *Managing spatially variable Nitrogen: A Practical Approach*. Proceedings of the Annual ASEA Meeting, Michigan State University, paper No 99, p. 1142 (1999).

Graham, C., Van Es, H., Melkonian, J., and Laird, D. *Integrating model-based adaptive management of Nitrogen with Site-specific NIR-based carbon estimates in Maize production*. International Annual Meetings of American Society of Agronomy. Long Beach, California, USA, 107(2), pp. 1–2 (2010).

GRDC. *Precision Agriculture-fact sheet-How to put Precision Agriculture into practice*. Grain Research and Development Corporation, Kingston, Australia, pp. 1–6 (2010), Retrieved from www.grdc.com.au (January 1st, 2011).

Griffin, T. W., Popp, J. S., and Buland, D. V. Economics of Variable-rate applications of Phosphorus on a Rice and Soybean rotation in Arkansas. *Proceedings of the 5th International Conference on Precision Agriculture and other Resource Management*. Bloomington, Minnesota, USA, pp. 23–40 (2000).

Hammond, M. W. and Mulla, D. J. J. *Development of Management maps for spatially variable Soil fertility*. Proceedings of the 39th Annual Far West Regional Fertilizer Conference. Bozeman, Montana, pp. 36–38 (1988).

Haneklaus, S., Bloem, E., and Schnug, E. *Precision agriculture-New production system for an Old crop*, pp. 1–7 (2002), Retrieved from http://www.regional.org.au/au/gcric/2/10.htm (July 2nd, 2011).

Heiniger, R. W. and Meijer, A. M. Why Variable Rate Application of Lime has increased grower profits and acceptance of Precision Agriculture in the Southeast. *Proceedings of the 5th International Conference on Precision Agriculture and other Resource Management*. Bloomington, Minnesota, USA, pp. 576–659 (2000).

Jiyun, J. and Cheng, J. *Site Specific nutrient Management in China: IPNI-China program*. International Plant Nutrition Institute, Norcross, Georgia, USA, pp. 1–7 (2011), Retrieved from http://www.ipni.net/ppiweb/china.nsf/$webindex/27D05B2887D6B7EE482573AE0029344 8?opendocument.

Karlen, D. L. Andrews, S., Colvin, T. S., Jaynes, D. B., and Berry, E. C. Spatial and temporal variability in corn growth, development and yield. In *Proceedings of 4th International Conference on Precision Agriculture*. American Society of Agronomy, Madison, Wisconsin, USA, pp. 101–112 (1998).

Khakural, B. R., Robert, P. C., and Mulla, D. T. Relating Corn and/or Soybean yield to variability in Soil and Landscape characteristics. In *Proceedings of the 3rd International Conference on Precision Agriculture*. P. C. Roberts (Ed.). American Society of Agronomy, Minneapolis, Minnesota, USA, pp. 117–128 (1996).

Kitchen, N., Hughes, D. F., Sudduth, K. A., and Birrel, S. J. *Comparison of Variable rate to Single rate Nitrogen fertilizer application: Corn Production and Residual soil NO_3-N*. Proceedings of the International Conference on Site Specific Management of Agricultural Systems, Minneapolis, MN, USA, pp. 427–439 (1994).

Kitchen, N., Shahanan, J., Roberts, D., Scharf, P., Ferguson, R., and Adamchuk, V. Economic and Environmental Benefits of Canopy Sensing for Variable rate-N Corn fertilization. In *Proceedings of the American Society of Agricultural and Biological Engineers Annual International Meetings*. Reno, Nevada, USA, pp. 1–3 (2009), Retrieved from http://asae.frymulti.com/abstract.asp?aid=27259&t=1. (December 15th, 2010).

Krishna, K. R. *Agrosphere: Nutrient Dynamcis, Ecology, and Productivity*. Science Publsishers, Enfield, New Hampshire, USA, p. 453 (2003).

Krishna, K. R. *Peanut Agroeosystem: Nutrient Dynamcis and Productivty*. Alpha Science Internationl Inc. Oxford, England, p. 292 (2008).

Krishna, K. R. Agroecosystems of South India. *Nutrient Dynamics, Ecology and Productivity*. Brown Walker Press Inc, Boca Raton, Florida, USA, p. 553 (2010).

Krishna, K. R. *Maize Agroecosystem: Nutrient Dynamcis and productivity*. Apple Academic Press Inc, Toronto, Canada, p. 325 (2011).

Kitchen, N. R., Sudduth, K. A., Drummond, S. T., Scharf, P. C., Palm, H. l., Roberts, D. F., and Vories, E. D. Ground-based Canopy reflectance sensing for Variable rate Corn fertilization. *Agronomy Journal*, **102**, 71–82 (2010).

Kyveryga, P., Caragea, P., Kaiser, M., Nordman, D., and Blackmer, T. *The dilemma of reducing Nitrogen fertilizer rate for Corn: Data analysis of On-farm trials*. International Annual Meetings of American Society of Agronomy. Long Beach, California, USA, 316(3) pp. 1–2 (2010b).

Lowenberg-DeBoer, J. and Reetz, H. F. Phosphorus and Potassium Economics for the 21st Century. *Better Crops*, **86**, 4–8 (2002).

Maine, N., Nell, W. T., Alemu, Z. G., and Barker, C. Economic analysis of Nitrogen and Phosphorus application under Variable and Whole field strategies in the Bothaville district of South Africa. *Research Report of Department of Geography*. University of Free State, Bloemfontein, South Africa (2005), Retrieved from http://ideas.repec.org/a/ags/agreko/7047.html pp 10.

Maine, N., Nell, W. T., Loweberg-DeBoer, J., and Alemu, Z. G. Economic Analysis of Phosphorus Applications under Variable and Single rate Applications in the Bothaville District. *Agrekon*, **46**, 72–78 (2007).

Mallarino, A. P. and Wittry, D. J., Dousa, D., and Hinz, P. N. Variable rate Phosphorus Fertilization: On-farm Research methods and Evaluation for Corn and Soybean. *Proceedings of the 4th International Conference on Precision Agriculture*. American Society of Agronomy, Madison, WI, pp. 687–696 (1998).

Mayer-Aurich, A., Gandorfer, M., and Wagner, P. Economic potential of Site Specific Management of Wheat production with respect to grain quality, pp. 1–6 (2007), Retrieved from www.efita.net/apps/accesbase/bindocload.asp?d=6261&t=0 (May 28th, 2011)

Miao, Y., Mulla, D. J., Robert, P. C., and Hernandez, J. A. Within field Variations in Corn yield and Grain quality responses to Nitrogen fertilization and Hybrid selection. *Agronomy Journal*, **98**, 129–140 (2006).

Molin, J. P., Motomiya, A. V. A., Frasson, F. R., Faulin, G. C., and Tosta, W. Test procedure Variable-rate fertilizer on coffee. *Acta Sciencia Agronomy*, **32**, 1–13 (2010).

Moshia, M. S., Khosla, R., Davis, J. G., Westfall, D., and Reich, R. *Precision Manure Management across Site Specific Management Zones*. International Annual Meetings of American Society of Agronomy. Long Beach, California, USA, 102(1), pp. 1–2 (January 11th, 2011) (2010).

Roberts, D., Kitchen, N., Sudduth, K., Scott, D., and Scharf, P. Economic and Environmental implications of Sensor-based N management. *Better Crops with Plant Food*, **94**, 4–6 (2009).

Ruffo, M. L., Weibers, M., and Below, F. W. Optimization of Corn grain composition with Variable rate Nitrogen fertilization. *18th World Congress on Soil Science*. Philadelphia, Pennsylvania, USA, pp. 1–3 (2006), Retrieved from http://a-c-s.confex.com/crops/wc2006/techprogram/P18257 (March 1st, 2011).

Runge, E. C. A. and Hons, F. H. Precision Agriculture: Development of a hierarchy of variables influencing crop yields. In *Proceedings of 4th International Conference on Precision Agriculture*. P. C. Roberts (Ed.). American Society of Agronomy, Madison, WI, USA, pp. 143–158 (1998).

Satyanarayana, T., Majumdar, K., and Biradar, D. P. New approaches and tools for Site Specific Nutrient Management with reference to Potassium. *Karnataka Journal of Agricultural Sciences*, **24**, 86–90 (2011).

Sawyer, J. E. Mallarino, A. P., Kiljorn, R., and Barnhart, S. K. *A General guide for Crop Nutrient and Lime stone recommendations in Iowa*. Iowa State University Extension Bulletin, pp. 1–18 (2008).

Scharf, P., Shannon, K., Sudduth, K., Kitchen, N., and Scott, D. Crop Sensors to control Variable-N rate N applications: Five years of On-Farm Demonstrations. *Proceedings of American Society of Agronomy Annual meetings*. Abstract, Pittsburgh, PA, USA, pp. 1–2 (2009a), Retrieved from www.ars.usda.gov/research/publications/publications.htm?seq_no_115=240682 (December 15th, 2010)

Scharf, P., Kitchen, N., and Bronson, K. Precision Nitrogen Management approaches to minimize impacts. In: *Proceedings of the American Society of Agronomy Annual meetings* (abstract) Pittsburgh Pennsylvania, pp. 1–2, (2009b), Retrieved from www.ars.usda.gov/resarch/publications/publications.htm?seq_no_115=240687 (May 28th, 2011).

Schepers, J. S. *Precision Farming: One key to Quality Water*, pp. 1–3 (1996), Retrieved from www.fluidfertilizer.com/pastart/pdf/13P28-31.pdf (May 28th, 2011)

Schloter, M., Bach, H. J., Metz, S., Sehy, U., and Much, J. C. Influence of Precision Farming on the Microbial community structure and functions in Nitrogen turnover. *Agriculture, Ecosystems, and Environment*, **98**, 295–304 (2003).

Simmonds, H., Linquist, G., Plant, R., and Van Kessel, C. The influence of flood irrigation in Rice systems on the Spatial Distribution of Soil nutrients, Plant nutrient uptake and Yield. *International Annual Meetings, Agronomy Society of America*, **106**(4) 1 (2010).

Singh, M. V. Evaluation of Micronutrient stocks in different Agro-ecological regions of India for sustainable crop production. *Fertilizer News*, **46**, 13–35 (2001).

Snyder, D. S. An economic analysis of Variable nitrogen management. In *Proceedings of the Third International Conference on Precision Agriculture*. P. C. Robert, R. H. Rust, and W. E. Larson (Eds.). Minneapolis, Minnesota, USA, pp. 1009–1018 (1996).

Sparovek, G. and Schnug, A. Soil tillage and Precision Agriculture: A theoretical case study of soil erosion in Brazilian Sugarcane production. *Soil and tillage*, **61**, 47–54 (2001).

Sudduth, K. A., Hummel, J. W., and Funk, R. C. Soil Organic Matter sensing for Precision herbicide application. *Proceedings of Conference on Pesticide formulations and application systems: Tenth symposium Philadelphia*, USA, pp. 111–125 (1990).

Sudduth, K., Newell, K., and Scott, D. Comparison of three Canopy Reflectance Sensors for Variable-rate Nitrogen application in Corn. *Proceedings of the International Conference on the Precision Agriculture Abstract*, pp. 1–2 (2010), Retrieved from www.ars.usda.gov/pandp/people/people.htm?personid=1471.html (December 15th, 2010).

Tucker, D. P. H., Alva, A. K., Jackson, L. K., and Weaton, T. A. *Nutrition of Florida Citrus Trees*. University of Florida Extension Services Publication-169, pp. 1–39 (1995).

Vanek, V., Balik, J., Silha, J., and Cerny, J. *Spatial variability of Total Soil Nitrogen and Sulfaur content at two conventionally managed fields*, pp. 1–10 (2008), Retrieved from http://journals.uzpi.cz/publicfiles/02456.pdf (July 2nd, 2011).

Viglizzo, E. F., Pordomingo, A. J., Castro, M. G., Lertro, F. A., and Bernardos, J. N. Scale-dependent controls on Ecological functions in Agroeosystems of Argentina. *Agriculture, Ecosystems and Environment*, **101**, 39–51 (2004).

Walley, F., Pennock, D., Solohuh, M., and Hnatowich, G. *Precision Farming: Precisely What do we know?*, pp. 1–16 (1998), Retrieved from http://www.ssca.ca/conference/2000proceedings/walley.html (January 1st, 2001).

Wang, H., Jin, J., and Wang, B. Improvement of Soil Nutrient Management *via* information technology. *Better Crops*, **90**, 30–32 (2006).

Wells, K. L. and Dollarhide, J. E. *Precision Agriculture: The effect of Variable Rate fertilizer Application on Soil test values*, pp. 1–12 (1998), Retrieved from http://www.uky.edu/Ag/Agronomy/Extension/ssvl196.pdf (June 11th, 2011).

Wortmann, C. S., Ferguson, R. B. Hergert, G. W., and Shapiro, C. A. Use and Management of Micronutrient Fertilizers in Nebraska. *University of Nebraska-Lincoln, Extension Bulletin G*, **180**, 1–4 (2008).

Yu, M., Segarra, E., Watson, S., Li, H., and Lascano, R. J. Precision Farming Practices in irrigated Cotton production in the Texas High plains. *Proceedings of the Beltwide Cotton Conference Vol 1*, pp. 201–208 (2001).

4 Geographic and Economic Aspects of Precision Farming

CONTENTS

4.1 GEOGRAPHIC ASPECTS OF ADOPTION OF PRECISION FARMING

Precision farming has been adopted in several agricultural belts of the world. It is now a popular agronomic procedure that involves series of sophisticated instrumentation and is specific to a site or field. However, according to Swinton and Lowenberg-DeBoer (2005) adoption of precision farming in different continents has been uneven and scattered in certain geographic regions and more pronounced in certain other farming regions. Further, they suggest that not all aspects of precision farming are adopted in all the regions. Actually, there are five aspects to precision farming. They are grid sampling, soil analysis and mapping, global positioning systems, variable rate inputs (fertilizer spreading), and yield monitoring. Yield monitoring is sometimes tied with soil fertility (productivity) mapping. Let us mention a few examples. Swinton and Lowenberg-DeBoer (2005) have stated that, variable rate technique has not been accepted uniformly throughout in North American plains, despite recent advances in sensor and satellite techniques. The situation with yield monitoring is similar. Not all farming regions make use of them to prepare a productivity map. Within a single nation or agricultural region, use of yield monitors is uneven. In the Midwestern USA,

11% farms are using yield monitors to judge soil productivity but in Southeastern states only 1.1% farms have adopted yield monitors. According to Daberkow and McBride (1998), in USA, most of the early adopters of precision farming techniques were well-educated and full-time farmers. They generally cultivated large sized farms of over 200 ha. In addition, they usually cultivated high value crops like sugarbeet, cotton or cereals that were in demand. The pattern of adoption of precision farming has varied depending on the sub regions within North American farming zones. For example, in Tennessee, survey indicated that yield monitor was among the most common technology adopted by farmers. About 62% of counties in Tennessee had accepted GPS methods. Grid sampling was most commonly used soil sampling procedures in 30% of cropping zones of Tennessee. During past decade, 8% of farmers in Tennessee practiced precision farming in at least one form (Burton et al., 2000).

In North America, variable rate techniques have also been regularly adopted to spread weedicides and pesticides (Swinton and Lowenberg-DeBoer, 2005). Clearly, factors like geography, economic feasibility, and yield gains affect the extent to which precision farming is practiced. It consequently affects nutrient dynamics and crop productivity in a farm or a large expanse depending on the context.

In Argentina, yield monitors are being accepted and used rapidly in many of the farms located in the Pampas. However, the same farms do not seem to use variable rate applicators, because fertilizer supply rates are low and inherent variations in soil fertility are not that great to warrant use of costly equipments like computer guided variable applicators (Lowenberg-DeBoer, 1999; Norton and Swinton, 2001; Swinton and Lowenberg-DeBoer, 2005).

In Brazil, it is said that precision farming approaches fit better in Rio Grande du sol and Southern Brazil, where farms are richer in soil fertility plus crops are of high quality and productivity (Lowenberg-DeBoer and Griffin, 2006). Management induced soil variability seems pronounced in southern and eastern Brazil. Agricultural cropping is expansive in Cerrados of Brazil. Introduction of precision techniques can improve nutrient use efficiency and grain yield in the above areas. Therefore, precision farming can reduce cost on fertilizer input. It is believed that farmers in this region may reap better economic advantages by adopting precision farming.

In Malaysia, site-specific methods are getting accepted rapidly in rubber plantations but not so in rice fields. In India, hand-held sensors and remote sensing data are being utilized to judge fertilizer need and crop yield, but variable rate instruments mounted on tractors or trucks are not common yet. According to Dobermann et al. (2004) adoption of precision farming in different geographical regions or specific agroecosystems is primarily driven by natural resources like soil fertility and its variability, cropping systems, exact demand for improvised technology, and economic advantages. For example, precision farming has been adopted to enhance fertilizer efficiency. In the North American Great Plains region and Australia, emphasis is on increasing profitability of farming enterprise through precision farming. To a great extent, precision farming has been adopted to reduce on fertilizer-N consumption and thereby reduce soil N accumulation and ground water contamination. As stated earlier, in most parts of Asian dry lands, precision farming has been accepted as a method to revise resource allocation and grain/vegetable yield goals. In most cases, precision farming allows farmers to revise their yield goals upwards and at the same time, obtain a certain degree of environmental advantages. The utility of precision farming is enormous within rice agroecosystems of Asia. It allows them to Taylor nutrient supply schedules based on exact needs of the crop at different stages. It also allows them to revise yield goals appropriately. In the Asian wetlands, it helps in preserving various natural processes, soil fertility and nutrient dynamics during rice culture.

4.1.1 Precision Farming in North America

Most important motives behind spread of precision farming techniques are to manipulate nutrient dynamics, to achieve uniformity in soil fertility, improve productivity, and overcome economic pressure. In North America, there are a number companies and private agencies that cater to a large population of farmers who adopt precision farming. Currently, most combine harvesters are fitted with devises that provide yield maps. Variable rate applicators are also common in large commercial farms (Blackmore, 2003). During mid 1990s, crop production experts evaluated major crops like wheat, corn, barley, soybean, and potato grown in Northern America for response to variable rate fertilizer supply. Grid sampling was carried out in majority of the experiments and fertilizers managed included combinations like, N:P:K, N:P and P:K. Lowenberg-DeBoer (1995) has opined that initially there were large number of field trials that provided better fertilizer efficiency. Soil nutrient dynamics could be managed better using VRT. The VRT required lower quantities of fertilizers, yet, it resulted in grain yield levels similar to conventional or uniform-rate application. The intensity of grid sampling and depth of soil profile examined affected crop response to VRT. Intensive grid sampling provided soil maps depicting soil fertility variations with better resolution. Currently, there are indeed many regions within North America or elsewhere in other continents, where in a wide range of crops have been tested for profitability, if cultivated using precision farming. There are also innumerable reports on influence of precision farming approaches on dynamics of nutrients in a single field, a farm or even large cropping zones. Although not exhaustive a few examples have been discussed briefly. They cover different geographic regions, soils and environmental conditions experienced in North America.

As stated earlier, in North America, major thrust is to achieve uniform soil N distribution and economize on fertilizer-N inputs. Let us discuss a few examples. Wheat producers in Oklahoma have been exposed to advantages of using sensors to supply nitrogen. A comparative study by Arnall et al. (2010) has shown that variable rate N application allows farmers to reduce N inputs to wheat fields. Fertilizer-N was applied to N rich strips based on sensor (N-rate calculator) reading. Sensor-based technique actually reduced N supply by 22 kg N ha^{-1}. Farmer's normal practices had envisaged 65 kg N ha^{-1}, but precision farming required only 43 kg N ha^{-1}. Yet both the agronomic procedures resulted in 3.4 t grain ha^{-1}. Since the crop was on N rich strips, undue N accumulation in soil profile was effectively delayed, if any.

Wheat production in Northwest USA is carried out on undulated landscapes. Commonly adopted crop rotations include spring lentils, winter wheat and spring peas. It is said that fertilizer-N inputs are crucial for optimum wheat grain productivity. During recent years, precision farming techniques that allow accurate distribution of fertilizer-N and other nutrients have been adopted for wheat production in this region. According to Fiez et al. (1994), spatially variable N supply based on yield goals is more economical than single rate supply. The cost of crop production using precision farming is relatively higher because of instrumentation and skilled labor requirements. However, this aspect is usually taken care of due to the reduction in fertilizer-N costs under variable rate N supply.

Sudduth et al. (2007; 2010) have evaluated a series of commercial crop canopy reflectance sensors with a view to rapidly assess plant-N status, then, feed the information electronically to variable N applicators, so that fertilizer-N dispensed into field is most uniform. Obviously, basic intention is to enhance fertilizer-N use efficiency. At the same time, it reduces loss of N to lower profiles of soil, avoids unnecessary accumulation or loss *via* emissions. In regions with soil N dearth, it allows for optimizing N availability, so that crop productivity is not reduced. Some of the commercial sensors tested were GreenSeeker, CropCircle, and CropSpec. The fertilizer-N application systems worked well. It created uniform availability of soil N to roots. However, we should realize that factors such as soil moisture, pH and other physico-chemical traits too affect N availability and absorption rates. We also need comparable control plots that allow us to judge the effects of sensors and variable applicators on fertilizer-N distribution and use efficiency. Sudduth et al. (2007) have reported that response of maize crop treated with fertilizer-N based on sensors (reflectance) and variable applicators turned out to be profitable. However, soil N versus moisture interaction too plays a role in N dynamics per square and economic advantages.

Kent et al. (2007) have pointed that it is routine to prepare a non-nitrogen limiting strip to compare crop reflectance and arrive at accurate estimates of crop N status and fertilizer-N needs. Multiple strips have also been used to standardize crop reflectance measurements and decipher spatial variability. Kent et al. (2007) compared alternative methods to using reference strips. One of the approaches is to earmark a small area within a field to determine a non-nitrogen limiting crop reflectance value that could be used to guide fertilizer-N input into the field. Another approach is to use electrical conductivity (EC) as parameter that could be correlated with reference strip to decide fertilizer-N inputs. The idea is to develop N input schedules to match crop growth and yield formation. However, we should also realize that accuracy of sensor reading, reference strips and models that decide on variable fertilizer-N inputs are some of the most crucial aspects. They have direct impact on N dynamics in the maize field. No doubt, instrumentation and techniques have their own particular influence on N input sizes. Fertilizer-N supplied using a particular method may impart greater influence on N cycle in a field, entire farm or area, where precision farming is practiced for example in the Corn Belt of USA. Hence, farmers need to be cautious in selecting appropriate techniques, simulations and computer models, while deciding N supply to maize.

When extrapolated to entire Corn Belt, influence of precision farming on N dynamics could be substantial. Following example depicts influence of precision farming on N dynamics, especially N supply and use efficiency, within the Corn Belt of USA.

Location in USA	Farming System kg ha^{-1}	Nitrogen Input t ha^{-1}	Grain Yield kg Grain kg^{-1} N	Nitrogen Use Efficiency
Lincoln, NE,	Conventional	142	10.3	73
	SSNM	113	10.2	90
Fort Collins, CO	Conventional	152	12.8	84
	SSNM	109	12.9	18

Source: Dobermann et al. 2004.

Precision farming practices have been evaluated across different locations in southern and central regions of Great Plains in USA. Currently, precision farming is more routinely used on cotton grown in southern high plains of Texas, since it reduces on fertilizer and irrigation costs. It also provides uniformity in terms of soil fertility and lint yield. Yu et al. (2001), argue that precision farming is preferable in the cotton belt because of environmental concerns related to excessive fertilizer and chemical inputs. Further, we should note that net economic gains range from 30–38 US \$ ha^{-1} due to precision farming compared to conventional farming. During this past decade, popularity of variable rate fertilizer supply methods has grown rapidly in Texas. Surveys have clearly shown that farmers use GPS guided systems and previously prepared soil fertility maps to distribute major nutrients to cotton fields. Most farmers gain both in terms of fertilizer saved and monitory advantages (Banerjee and Martin, 2007).

Roberts et al. (2002) too state that aspects like extra profitability, reduction in fertilizer and chemical inputs into fields, safe guard of soil quality and agro-environment have been considered priority items by cotton farmers of USA. Cotton farmers in southern states of USA adopt at least one or more of the precision farming techniques. Grid and management zones are most popular methods to demarcate fields based on soil nutrient status or other characteristics. Precision farming is mostly adopted to apply variable rates of N, P, K, and lime. According to Lambert and Lowenberg-DeBoer (2000) and Roberts et al. (2002), cotton farmers preferred precision farming since it was profitable and reduced fertilizer inputs into their fields. Here, crop value seems to be the most important determinant of profitability of precision farming. It is interesting to note that farmers with better managerial skills and education driven by advantages in terms of soil nutrient dynamics, environment and profitability, adopted precision farming more easily.

Rice production is feeble in North America. It is not a major cereal in USA, yet wetlands in some southern states allow paddy cultivation. According to Tran and Nguyen (2008) small number of rice farmers situated in North America, especially in the states of California, Florida, and Louisiana do adopt precision farming techniques. They stringently follow most of the procedures that improve production efficiency. For example, they adopt grid sampling, computer-aided simulation models and decision support systems, variable rate applicators, sensors, GPS aided crop growth monitoring, harvest machinery equipped with yield monitoring and recording devices, post harvest machinery that electronically grade harvests and finally computers that record detailed data about all farm operations, inputs and harvests. Yet, there are several constraints specific to American farmers that curtail or delay use of precision farming procedures on rice.

They are:

(a) Gathering accurate information regarding soil nutrient distribution and other factors that affect crop growth and grain formation is expensive and time consuming;

(b) Exact benefits from precision farming, especially those relating to soil quality and agro-environment of wet lands are not perceivable immediately.

It may require several crop seasons before it is conspicuous enough. No doubt, costs on use of precision farming methods are decreasing. Yet, in many instances they are not appropriately low. Lack of skilled personnel, who can handle large amount of computer-based data and GPS guided tractors make this technique costly (Tran and Nguyen, 2008).

Potato farming is an important economic activity in North and Northwest of USA, especially in the states of Idaho, Montana, North Dakota, Washington, and so on. In addition to N, P, and K, application of FYM is necessary, if crop yield goal has to be met. Currently, several farmers adopt precision farming methods. Such methods include soil survey, soil test for nutrient elements, soil fertility maps to identify variations, computer-aided decision support systems and variable rate nutrient inputs. In case of P and K, Hammond (1993) found that preparation of soil fertility maps and adoption of VRT costs more than conventional fertilizer supply methods. Yet, VRT was preferred because it resulted in good quality harvest. In Central Washington, net profit from VRT was 75–150 US\$ ha^{-1} more than conventional method.

In the sugarbeet cultivating zones around Great Lakes region, profitability from precision farming seems to depend immensely on net variability in soil fertility. Grid sampling should indicate significant variation in nutrient distribution, both in surface and sub-surface layer of soil, if precision farming has to be profitable. Soil nitrogen $<$ 33.0 kg ha^{-1} provided better returns on N inputs *via* VRT. It is interesting to note that 70% of area supporting sugarbeet has enough variation in soil fertility to warrant use of VRT (Lilleboe, 1996). Forecasts indicate that in about 80% of the locations examined, sugarbeet treated with fertilizer-N using VRT could result in profits. It is said that weather pattern during the season may also affect N dynamics leading to alterations in net returns from VRT.

4.1.2 Precision Farming in South American Cropping Zones

According to Lowenberg-DeBoer and Griffin (2006) precision farming that includes grid sampling and testing of soil, mapping fertility variations, computer-based decision support systems and variable rate fertilizer is a promising technique for South American farmers. There are several aspects of precision farming that suit Brazilian agriculture. Let us consider a few of those aspects. As stated earlier, precision farming suits the cropping trends in Southern and Eastern Brazil, where land cost is high, inputs into farm is high and important crops like maize, legume, soybean, and coffee are cultivated. In the Cerrados, where cropping is expansive, precision farming may find easy acceptance. It may provide farmers with greater economic advantage. The costs of instrumentation, skills and training may slow down spread of precision farming. Although, GPS guided systems do save on labor costs, it may not make a dent as for as advantages from precision farming is concerned. Yet, in many locations, it has been observed that availability of labor and its cost is not a constraint. Lowenberg-DeBoer and Griffin (2006) point out that Cerrados is a region with low fertility acidic soils. Farmers adopting precision farming should first correct the soil pH using variable lime inputs. Later, supply major and micronutrients in appropriate quantities. Precision farming could be most useful in correcting nutrient imbalances, soil reaction, and still be economically highly viable. Although, precision techniques aim at a single field, we should realize that upon extrapolation, its impact on nutrient dynamics and soil productivity in the entire Cerrados could be immense. Precision agriculture may actually help Brazilian farmers by reducing fertilizer supply to soil and cost on inputs. It is believed that as such, precision techniques could be most useful in regions supporting cereals, soybean and pastures. It could also be highly profitable in citrus groves, sugarcane fields and rubber plantations.

Precision farming in Brazilian cropping zones has been focused more towards avoiding over or under application of fertilizers, irrigation and pesticides. It is also aimed at correcting soil and nutrient loss due to erosion. In the Percicaba, precision techniques are being evaluated on cereals, legumes and sugarcane. Reports suggest that soils (Dystrochrepts, Paleudults and Udorthents) vary significantly for nutrients and several other physico-chemical properties even within small areas in a field. Large uniform applications of fertilizers or mechanical operations may not remove variations. Hence, adoption of precision farming seems pertinent. Sugarcane is a major crop in Brazil. According to Sparovek and Schnug (2001) precision farming could be useful in regulating soil erosion and nutrient loss, obtaining uniformity soil P availability and improving economic efficiency of sugarcane production in the southeastern plains of Brazil.

Soybean is a major legume in the Cerrados of Brazil (Cerrados). Its productivity is highly dependent on soil physicochemical traits and fertilizer inputs. Soil amendments like lime is important to overcome Al toxicity and pH related problems. Grain yield variability is high ranging from 2,100–6,800 kg ha^{-1} (Bernardi et al., 2003). Soil maps show equally high variations with regard to nutrients. Precision techniques seem to overcome many of the soil related problems and provide farmers with uniform productivity. Most importantly, precision technique helps the farmers in avoiding over and under supply of nutrients, that otherwise occurs under conventional methods.

According to Bongiovanni and Lowenberg-DeBoer (2005), Argentinean farmers aim firstly to reduce costs incurred on farming and grain production. Then, they try to increase productivity per unit area. Farmers actually aim at more efficient use of fertilizers. Further, they have opined that, although precision technique does nothing to affect aspects like transportation of grains, credits or pricing policies of the market, still Argentinean farmers intend to stay aloft in the world grain market by reducing the cost of cereal grain production. They hope to achieve greater efficiency of fertilizer and irrigation by adopting precision techniques. Currently, there are over 20 large agricultural companies that supply precision equipments and computer-based packages to farmers in Pampas. Soil fertility mapping is an important step during precision farming. Yet, only 50% of Argentinean farms are regularly tested for soil nutrients. About 25% farms adopt soil testing just prior to planting and only a small fraction (2%) resort to intensive soil sampling and analysis. It seems only a few farms adopt variable rate technology (VRT) to supply fertilizers. The VRTs are confined to application of fertilizer-N. Such VRTs are mounted on trucks or tractors. The commands for VRTs are derived from short distance handheld sensors, "on-the-go" sensors and remote sensing. Data from previously prepared soil fertility maps are also used to guide VRTs during nutrient supply.

There are indeed several constraints to rapid spread of precision farming approaches in Argentina (Bongiovanni and Lowenberg-DeBoer, 2005).

A few of them quoted often are:

(a) Higher initial investment on equipments and training,
(b) Risk due to fluctuations in interest rates, grain harvests and market price,
(c) Less soil variability caused due to farming,
(d) Wide spread custom operators for farming operation, and
(e) Lack of enthusiasm to adopt high input technology, since low cost techniques are performing better.

Farmers usually weigh the economic advantages and ease with which a technology could be adopted. Lowenberg-DeBoer (2000) mentions that precision farming was still rudimentary at the turn of the century in 1999 A.D. The usage of combine harvesters with facility to monitor yield was low at 1–2% compared with 4–8% at present. It was attributed to lack of high variability caused due to cropping and agronomic procedures. The soil fertility variation that existed was mostly due to natural factors. Farmers applied only marginal levels of fertilizers at 30–60 kg N, 20 kg P, and 20 kg K ha^{-1}. This did not allow any great margin to save on fertilizers through precision farming. As a consequence, variable rate applicators were feebly used. Many of the farms in Pampas were built on marginal soils with low productivity and were also relatively small in area. They were less than 4,000 ha. It is clear that economic factors are important constraints to adoption of GPS guided precision farming in Argentinean Pampas. During 2,000 A.D., factors like prevailing demand for corn, grain productivity, and economic incentives did not allow rapid spread of precision farming. Precision farming using GPS guided traction was adopted mostly to spray pesticides. Following data regarding use of equipments related to precision farming is indicative of growth and acceptance of precision farming techniques in Argentina:

Precision Equipment	**Year**	
	1997	2005
Yield Monitors	25	850
Variable Rate Fertilizer Applicators	3	40
GPS Guided Systems	35	3053
On-the-go Nitrogen Sensors	0	7

Source: Bongiovanni and Lowenberg-Deboer, 2005

Perhaps, the number of equipments (sets) relevant to precision farming in the Argentinean farming zones may have now more than doubled since 2005 levels.

Chilean farmers utilize precision farming to regulate N dynamics in the fields that support production of wheat, maize and oats. They intend to reduce fertilizer-N supply and remove variations in soil N availability. It is said that Rhodoxeralfs found in Chilean farming zones exhibit rampant variations in soil N and SOM content. Soil N mineralization rate too varies conspicuously. Precision farming approaches are also directed to reduce loss of N *via* leaching to subsurface layers and ground water. At present, Chilean farmers who practice precision farming, apply about 30–40 kg N ha^{-1} less than ones who depend on traditional farmers practice (Claret et al., 2011).

4.1.3 Precision Farming in the European Farming Belts

Precision farming and GPS guided grain harvests have been in vogue in the European plains for the past 15–20 years (Pedersen, 2011). Precision farming approaches may find preferential acceptance by European farmers due to several reasons. Prescription based on individual field is an important and attractive proposition to farmers. In addition, use of soil maps that specify fertility, moisture, and pH all seem very useful. Sensors, especially "on-the-go" sensors and VRT are supposedly the main reasons for the modest popularity of precision technology (Lowenberg-DeBoer, 2003b; Sylvester-Bradley et al., 1999). Farmers in Europe impinge their fields with relatively high levels of major nutrients N, P, and K, totaling 275–300 kg fertilizers ha^{-1} $season^{-1}$. In addition, micronutrients are added to obtain proper ratios and nutrient balance in the soil ecosystem. Obviously, high fertilizer supply is to ensure greater productivity of cereals. However, environmental impact of high fertilizer and other soil amendments seems really adverse. Therefore, in the Western and Northern European region, state agencies regulate supply of N to soil. In Denmark, Germany and United Kingdom, fertilizer-N is rationed and provided to farmers on a quota basis. This is to guard against further deterioration of soil. Precision farming techniques that maintain grain yield levels yet provide greater N use efficiency are a boon to farmers in Western European plains (Dobermann et al., 2004). Improvement in N use efficiency may be attributed to correct timing, exact dosages, better placement near the roots, appropriate nutrient ratios (e.g. N:P, N:K, N:C etc.), reduction in N loss *via* erosion and emissions and so on. Application of fertilizer-N using variable applicators improved cereal production efficiency by 12–52%. It actually helped in reducing soil N accumulation. Reduction of surplus N in soil has its advantages. It lessens risk of loss *via* erosion and emissions. It also reduces chances of groundwater contamination.

Distribution of fertilizer-N in wheat fields has been attempted using plant and soil sensors as well as data on topography (Berntsen et al., 2006). Initially, fields are divided into subplots and fertilizer-N supplied to achieve uniform soil N availability. European farmers have evaluated several computer models to direct variable-N applicators. It seems plant-N sensors predicted fertilizer-N requirement and grain yield better than soil N sensors. Further, methods based on plant-N sensors have clearly shown that N should be amended, so that zones with low or high N concentration get equilibrated to moderate levels of soil N.

As stated, wheat is an important cereal crop in the Western European plains. It is intensively cultivated with high dosages of fertilizers and irrigation. Soils are highly variable in terms of fertility, reaction, and grain yield. Hence, precision techniques were sought to improve fertilizer efficiency and grain productivity. Currently, wheat farms and cooperatives in France and other regions routinely obtain digital imagery from remote satellites (e.g. SPOT). Remote sensed images provide detailed information on soil fertility variations, crop status, and yield prediction. Commercial use of SPOT derived imagery of wheat fields began in 2002, after 6 years of standardization and experimentation in the plains (Astrium, 2002). The procedure involved is fairly simple. The farm cooperatives and farmers receive pictures of their fields (soil and or crop) routinely with detailed notes on soil fertility, nutrient dearth, fertilizer recommendation, and crop status. In-season pictures are also supplied so that accurate decisions regarding split application of fertilizer-N is possible. Yield gains due to adoption of satellite derived digital imagery and decision support system was 1.9 sq ha^{-1} (Astrium, 2002). Perhaps, this procedure will soon be adopted in wider regions of Europe.

Precision farming approaches have been evaluated on paddy (*Oryza sativa*) produced on sandy loams found in southwest of Portugal. Three years of field tests at Salso farm in Vitoria involved extensive soil sampling to prepare soil fertility maps, demarcation of fields into management zones and application of fertilizer based nutrients N, P, and K, using variable rate techniques. In comparison to whole farm supply of fertilizers, site-specific methods increased soil fertility and provided better uniformity. Paddy grain/forage yield maps indicated that crop productivity was uniform. Adoption of management zones decreased soil fertility variability. However, there was no correlation between precision farming techniques and rice grain quality in terms of seed weight and protein content.

Forecast by researchers suggest that precision farming techniques may find their way into areas supporting viticulture, especially in southern European regions. Precision techniques are said to be most useful in fine-tuning the fertilizer and pesticide use in the vineyards. Precision viticulture seems viable because of the value of the crop

(Lowenberg-DeBoer, 2003b). Overall, spread of precision techniques in European farming zones seems rapid.

4.1.4 Precision Farming in Africa and West Asia

Precision farming is a fairly recent introduction into Southern African cropping zones. Of course, its utility is being evaluated in many countries of this region. Its greatest advantage lies in the fact that it reduces on farm inputs, especially fertilizer-based nutrients and irrigation to achieve almost similar grain/forage productivity. The other set of benefits relate to reductions in nutrient accumulation and groundwater pollution. According to Maine et al. (2005), fertilizer accounts for 25% of total input costs in this region. Reduction of costs incurred on fertilizer-N and improvement of fertilizer-N efficiency is indeed a major preoccupation. Therefore, resource poor farmers in Southern African farming belts prefer techniques such as site-specific nutrient management (SSNM) and variable rate application of nutrients.

Introduction of precision farming methods in South Africa has depended on several factors related to farmer participation, availability of historical data of the field; especially its cropping pattern, fertility status, nutrient distribution in the soil profile, equipments, computer skills, inputs and profitability. In South Africa, precision farming has been applied to evaluate tillage methods, fertilizer supply, crop varieties and seeding rates. Some of the early experiments using precision farming techniques were conducted in maize farms around Bloemfontein in Free State during 2003–2005. Precision farming techniques that involve N and P inputs based on soil mineral analysis is surely replacing the erstwhile standard rates of fertilizer supply. This is true with many field crops and a few horticultural crops. According to Matela (2001) variable rate of nutrient input is mainly concerned with N and lime that are important for cereal production in South Africa. It has been argued that a change in fertilizer quantity and method of application is mainly due to exorbitant rise in fertilizer costs that could be offset through precision farming. Maine et al. (2005) believe that profitability of precision farming is the most important criterion for farmers in South Africa. Initially, its adoption will revolve around fertilizer inputs into farms. Economic advantages will decide its future as a sought after technology among the farmers. Popularity of precision farming techniques and use may taper off, if profits are feeble or erratic. Maine and Nell (2005) suggested greater caution while adopting precision farming on horticultural cash crops in South Africa. They hinted at achieving at least 15% greater advantage in terms of profitability plus environmental benefits. They also suggest evaluation of various techniques and gadgets relevant to precision farming to fit to horticultural farms. In addition, a critical assessment of impact of precision guided fertilizer and chemical supply is required. Some of the strategies suggested and adopted in South African horticultural farms have involved soil analysis based on thorough grid sampling of the area in question; use of yield monitors on a long term basis to judge soil/field productivity; fertilizers and lime supply using variable rate applicators and so on. A full fledged package of precision farming meant for horticultural crops grown in South Africa has always included grid sampling, remote sensing data, sensors for soil analysis, yield monitors and GPS guided variable rate applicators.

Jhoty and Autrey (2000) have suggested that adoption of precision farming techniques in the sugarcane belt of Mauritius could lower cost of production. It may reduce fertilizer inputs and undue accumulation of nutrients in the subsurface layers of soil. There is a strong need to assess long term effects of precision farming on soil and environment within the sugarcane cropping zones of Mauritius. Precision farming may actually allow us to regulate nutrient dynamics more stringently within the sugarcane fields and cropping expanses of Mauritius.

Precision farming is in vogue in Middle east Asia. Date palm is an important cash crop in many of the Arab countries. It is grown in well layer out plantations. Blocks are managed using previous soil fertility and yield data. Fertilizers are applied manually both in traditional and in precision plots. Yield maps are prepared by recording dates harvested from each tree. Since much of the steps are manual cost of adopting precision is relatively small (Blackmore, 2003).

4.1.5 Precision Farming in Asian Cropping Zones

Precision farming has been in vogue in different forms within Asian farming belts, since two decades. The extent of technological improvements adopted, accuracies achieved with regard to soil fertility management, crop produce levels and actual economic or environmental advantages accrued have varied. Major field crops of relevance to Asian farming zones that are discussed within this section are rice, maize, sorghum, legumes such as pigeon pea, soybean, oilseeds like groundnut, sunflower, mustard, tuber crops like potato, fibers like cotton, and so on. Horticultural crops like mangoes and citrus have also been cultivated using precision techniques. Let us consider a few examples.

Precision farming was introduced in China during early part of 1990s. Research and adaptation of precision techniques was actually aimed to address different aspects of Chinese agriculture and natural resources. For example, it aimed at solving problems related to fertilizer supply, crop production, forest management, ecological balance, and economic benefits. Precision techniques could be classified based on several different criteria. Based on tools used it was categorized into traditional technology, information technology based semi-mechanization, and highly automatic satellite-based technique (Jintong et al., 2010).

For small-scale maize or wheat farms of Shanxi province of China, precision farming or site-specific farming is a method that allows increased income. The small family plots are maintained by identifying soil variability and implementing rational nutrient application (Wang et al., 2006). Soil fertility variations mostly pertained to NH_4^+-N, P, K, and org-C. On poorly drained Fluvo-aquic soils, adoption of SSNM improved maize grain yield by 5–25%. Precision farming techniques are being routinely used during rice production in Southern China. Experiments and field demonstrations have shown that adoption of grid sampling, sensors, soil fertility maps, decision support systems, and variable rate applicators in the rice belt could be beneficial to farmers, in terms of nutrient dynamics within the field, ecology, and economic benefits (Xie et al., 2007). Reports by Jiyun and Cheng (2011) suggest that SSNM has been successfully adopted in several provinces of China, especially in areas that grow cotton, maize and wheat. In provinces like Hebei and Shandong, farmers cultivating maize/wheat have reaped higher grain yield, through the use of SSNM compared to Farmer's practice. The improvement in N dynamics, especially aspects like recovery from soil and higher fertilizer-N use efficiency (36–40%) has been the factor inducing spread of SSNM in Chinese farming zones. It is interesting to note that site-specific techniques have been repeatedly tested and adopted at various levels of farming like individual field, farm, village, province, and large expanses (Jiyun and Cheng, 2011).

Rice farmers in the Malaysian Peninsula are currently adopting precision farming techniques to improve soil nutrient dynamics and economic benefits. They apply major nutrients based on detailed soil sampling, analysis of major and micronutrients, pH, CEC, and organic matter. Kriged maps depicting soil fertility variations are routinely used during variable rate nutrient supply (Aishah et al., 2010). Results indicate that variations are conspicuous for major nutrients like N and P. About 30% of fields were low for soil N and 70% for available P. Soil pH is generally low at 5.0 and needs correction appropriately. Norasma et al. (2010) point out that paddy farmer in Southeast Asia, particularly those in Malaysia look for two-fold advantages from precision farming. They aim at reducing cost on fertilizer by lowering dosages. Reduction in chemical fertilizer supply has definite effect on soil nutrient accumulation, loss *via* erosion, emissions and recycling through Stover. Secondly, paddy farmers believe that for almost same amount of fertilizer supply grain/forage yield could be improved through adoption of site-specific techniques.

Rice is cultivated with reasonably high inputs of fertilizers, irrigation and pesticides. The productivity of hybrid rice is held high at 12–14 t grain ha^{-1} plus forage. Blanket recommendation of fertilizers by state agencies are indeed among the highest ranging from 180–230 kg N, 60–80 P, and 130–170 kg K ha^{-1}. Nitrogen is a key element whose dynamics has immediate effect on soil, ambient atmosphere as well as crop yield. Crop residue recycling procedures and fallows are crucial. Yet, fertilizer-N recovery rates are low at 30–40%. During recent years, precision techniques have improved fertilizer-N supply to rice fields. Soil tests, leaf color charts and chlorophyll meter readings have guided fertilizer-N supply. Fertilizer-N inputs have also been split into a small basal dose and in-season supply based on "on-the-go" sensor measurements. Soil maps based on EC measurements and soil sensors that estimate nutrient status have also been used to identify spatial variations. Dobermann et al. (2004) state that in the Asian rice belt, currently farmers use SSNM more frequently. It involves supply of variable rates of N, P, and K based on "on-the-go" chemical analysis, decision support systems based on improvised computer models and yield goals. This step reduces nutrient accumulation in soil profile. It avoids ground water contamination. Nutrient recycling *via* residues can be gauged accurately. Precision techniques improved fertilizer-N efficiency by 4% and grain yield improved by 11% over fields maintained under conventional systems.

Thailand is a major rice and legume producer among Asian nations. Commercial farms are maintained using large inputs of fertilizers. Crop yields are moderate in most regions. Hence, enhancing crop productivity and economizing on fertilizer-based nutrient is a preoccupation. Researchers and farmers have devised and tested wide range of techniques that reduce nutrient inputs and enhance crop productivity. Among them, precision farming is an important recently introduced technique. Precision farming is supposed to correct the soil nutrient dynamics. Incidentally, in Thailand, major soil related problems are inappropriate SOM concentration, nutrient imbalance and rampant nutrient removal. Nutrient loss due to soil erosion is also high. These factors have caused soil fertility variations (Hongprayoon, 2010). It is believed that adoption of precision farming will help in overcoming variations in soil nutrient availability. Precision farming has been tested on many crops like cereals, legumes, sugarcane, and cassava. In the western region of Thailand, precision farming adopted on sugarcane involves use of yield maps, DGPS, GIS, and variable rate applicators fitted on to tractors. Grid points that are closely placed help in deciphering nutrient availability. Series of field trials with maize during 2000–2005 has shown that precision farming helps in developing efficient fertilizer recommendations. The fertilizer supply depended on models used in the decision support system fitted to tractors. Variable rate applicators have been used to supply fertilizers like N, P, K, and gypsum to correct soil pH variations. Farmers reap better harvests based on yield goals fixed. Reduction in fertilizer supply and suppression of nutrient loss from ecosystem are other advantages. Actually, farmers producing maize adopt a decision support system called *SimCorn*. It is a computer-based simulation model. It provides fertilizer (N, P, and K) recommendations based on generally known N algorithms for fertilizer-N, simple P algorithms for fertilizer-P and Mitsherlich's equation for fertilizer-K. Precision farming consistently improved net profits derived from maize fields (Attanandana et al., 2000; Kasetsart University, 2006).

Paddy is an important cereal in the Fareastern Asian nations like Japan and Korea. It is grown intensively with high rates of fertilizer-based nutrient supply. Nutrient turnover rates in the farm and landscape are high. Farmers often supply 250–325 kg fertilizer ha^{-1} each season, to bolster high crop harvests ranging from 9–11 t grain ha^{-1} plus forage. Hence, Blackmore (2003) argues that, basic intent in using precision farming techniques in these Far eastern nations is to regulate nutrient dynamics and protect the agricultural environment. The economic benefits too are important. Farmers in Japan sell their produce to a protected market that pays them 5 times more than the general price prevalent in other agricultural areas. Hence, adoption of precision farming is generally remunerative.

Paddy farmers in Korea have been exposed to precision farming methods that use sensors to assess soil properties such as texture, pH, EC, moisture, micronutrients, total organic carbon and so on. Sensors that function at visible and near-infrared wavelength provide useful data about soil variability. Such data could be effectively channeled to variable rate applicators that are used to correct soil nutrient distribution, pH and moisture (Chung et al., 2008).

One of the main factors that induce farmers in India to use precision techniques is that it avoids excessive use of fertilizers and other inputs. It provides uniformity to soil fertility and stabilizes grain yield. It can be adopted in most areas and on several crops. Major concerns to Indian farmers are area coverage (location), size of farms/fields, data management, scale of farming enterprises; availability of infrastructure, especially soil maps, traction equipment and GPS systems; privacy of soil fertility and productivity data and finally economic viability of precision farming. Northwest India supports hardy cereal like pearl millet during kharif season and wheat during winter (rabi). They indeed constitute a major share of staple cereal food for the population in that region. Fertilizer supply is obviously higher than required in many fields. The net removal of nutrients by pearl millet and wheat that follows may not be commensurate with that impinged into fields. Hence, agronomic measures that impart greater fertilizer efficiency or those stipulating lower dosages like SSNM attract greater attention. Precision farming (or SSNM) that has been introduced has shown promise regarding reduction in fertilizer-N and P consumption (Dwivedi et al., 2011). The SSNM has a certain impact on several soil characteristics such as pH, nutrient availability and accumulation. Basically, SSNM imparts a degree of uniformity to soil nutrient distribution and crop harvests. In due course, as fertilizer-N and supply gets proportionately reduced, it might affect nutrient cycling in the vast area under wheat/pearl millet.

The SSNM techniques have also been experimented and adopted in some locations of Andhra Pradesh in South India. Here, arable crops thrive on soils that are poor in soil fertility. Variation in nutrient availability to crop plants is common. Sometimes crops suffer deficiencies of multiple nutrients. According to Srinivasa Rao (2011), adoption of SSNM proved useful in overcoming soil nutrient deficiencies more accurately. It also improved grain yield. In addition, SSNM provided protection to soil environment and enhanced benefit:cost ratio for major cereals grown in this region.

Precision farming based on soil chemical tests, crop sensors or remote sensed images and variable rate applicators are available for tuber crops like cassava grown in the Southern Indian hill regions. The decision support system uses QUEFTS model to recommend major nutrients N, P, and K. The QUEFTS model considers fertilizer requirements based on their availability to roots and yield goals set by farmers. Leaf color charts, chlorophyll meters and soil nutrient sensors are used to assess soil fertility and crop growth status periodically. It also considers nutrient interactions and ratios most congenial to cassava production (Byju and Suchitra, 2011). The SSNM techniques have also been tested on several other crops grown in the Central and Southern Indian plains (Blaise, 2006; Deshmukh, 2008; Mahavishnan et al., 2005). In case of cotton, Blaise (2006) believes that SSNM based on targeted yield should reduce nutrient loss from fields and enhance gross profits.

In Srilanka, tea is an important crop that is cultivated in well-marked blocks. A DGPS may not be needed if plots are well marked. Yield maps are prepared by recording tea leaf plucked from each block. Since inputs are applied manually, automatic variable applicators are not common on the plantations. The cost of implementation of precision farming techniques is relatively small (Blackmore, 2003).

4.1.6 Precision Farming in Australian Farming Zones

Precision farming has been in vogue in the Australian farming zones for more than a decade. It has helped in efficient production of cereals. In this region, precision farming techniques are mainly intended to optimize soil fertility traits, regulate nutrient dynamics and soil moisture status in a given location. Further, it helps in overcoming land and crop variability. Researchers at Grain Research and Development Center, at Kingston in Australia have summarized an array of crop production activities that precision farming approaches can accomplish, comparatively better than conventional techniques.

They are:

(a) Matching the seed rate to soil type to improve germination;
(b) Changing varieties or crops within a region to suit to the soil fertility conditions;
(c) Re-distributing fertilizers, especially N, P and K in a field or a region;
(d) Creating nutrient status and fertilizer input maps;
(e) Using crop biomass and N status maps;
(f) Inter-row weed control measures;
(g) Reducing soil compaction by regulating operation of traction equipment;
(h) Targeting within-season split N inputs to cereal crops as accurately as possible;
(i) Identifying most productive areas and preparation of accurate yield maps for focusing inputs and cropping sequences and
(j) Monitoring seed quality, especially seed protein content in a field location.

Since an array of soil characteristics, fertility and crop productivity are influenced by precision farming approaches, it is more than clear that a wide range of aspects of nutrient dynamics are affected. Firstly, nutrient applied to soil or a region reduces either marginally or appreciably, based on variations traced. Amount of nutrients absorbed and apportioned into grains or forage is influenced due to decisions provided by crop models and variable inputs. Aspects like biomass recycling, C sequestered and that lost may also be severely affected.

According to Bramely (2006), major assumptions that necessitate adoption of precision farming are: "land is variable" and then "no two soil particles, fields, farms or regions are the same". Research to improvise precision farming methods; dissemination of latest knowledge about techniques and their adoption in Australian agricultural regions is currently guided by "The Australian Center for Precision Agriculture". At present, this organization focuses on application of precision farming systems in Australia (ACPA, 2005). Aspects included are:

(a) *Spatial Referencing*: Collection of data on spatial variation of soil and crop features using geo-spatial positioning techniques.

(b) *Differential Action*: Response to spatial variations through altering farm operations such as tillage, sowing rate, fertilizer input rates, splitting and timing of nutrient supply, irrigation, and so on. variations in timing and intensity or quantity of farm inputs to match and coincide with known spatial variations.
(c) *Soil and Crop Monitoring*: Soil and crop features are monitored using various techniques such as chemical analysis, sensors, GPS using satellites, and so on.
(d) *Sensors or Satellites*: Prediction of Crop yield using Satellites and sensors.
(e) *Decision Support*: Knowledge of extent of variations in soil fertility and crop growth/production. Formulation of matching responses involves use of computer models and application of nutrient/water through variable rate applicators.

4.2 ECONOMIC ASPECTS OF PRECISION FARMING

There is no doubt that precision farming is a procedure that needs extra investment to support the grid sampling, soil analysis, mapping, sophisticated instruments like computers, variable rate applicators, skilled labor, and time. Generally, occurrence of small or large spatial variability in soil fertility and moisture distribution is considered a reason enough to resort to grid sampling, computer guided decision support and variable rate techniques. In addition, variable rate techniques supposedly lessen fertilizer/chemicals impinged into soil ecosystems. Hence, it imparts greater environmental protection. This apart, farmers save economically on fertilizer costs by adopting precision techniques compared to conventional or uniform fertilizer supply. One of the arguments in favor of precision farming is that fertilizers offset by precision farming pays up for extra investment on meticulous grid sampling, laboratory analysis of soil samples and computer guided variable rate application methods (Schepers, 1996). There are indeed several studies made in different continents that have aimed at understanding economic advantages of adopting precision farming. In addition to monetary gains, economic value of soil and environmental protection has also been evaluated in many cases. At the bottom line, for precision farming to be profitable following factors have to be considered. They are: (a) investments necessary for precision farming, (b) heterogeneity of location and soil fertility variations, (c) farm size, and (d) crop species and its economic value (Wagner, 1999).

We can try to ascertain each specific agronomic procedure that contributed to extra economic advantage when precision techniques are adopted. Firstly, precision techniques have been adopted to sow maize seeds. Lowenberg-DeBoer and Griffin, (2006) evaluated the cost of seeds plus variable rate seeding and compared it with conventional seeding systems. Seedling establishment and plant stand obviously has direct impact on grain yield. Field trials indicate that few extra seedlings planted under variable rate sowing affects profitability. Lowenberg-DeBoer and Griffin (2006) found that variable seeding potentially improves profits, if the soil is potentially low in fertility and yield levels are low. Variable seeding seems to be profitable, even if only 10% soil or field experiences low fertility in a farm of 400 ha (see Plate 1and 2).

Lowenberg-DeBoer and Swinton (1997) reviewed several research reports about economic advantages from precision farming. They traced all three possibilities. About 40% trials resulted in enhanced profits due to adoption of precision farming, 35% were not profitable and a few were ambiguous. The ambiguity in deciding the economic effects of precision farming occurs due to lack of standard criteria for comparison. Often, it is believed that response to major nutrients like N or P is highly predictable. Generally, profits from precision farming were significant, if value of the crop produce was high. For example, in case of cash crops like cotton, precision farming results in higher net profits. Accounting procedures also affect our interpretation about extent of profits from precision farming. Quite often, environmental effects are not equated properly and accounted in terms of cash benefits from precision farming. Since fertilizer-based nutrient inputs are dependent on economic considerations, indirectly such factors affect nutrient dynamics in the ecosystem.

Profitability of variable rate application (VRT) of fertilizers may vary based on geographic location, soil type, crop, and economic value of the grain/forage yield. For example, Swinton and Lowenberg-DeBoer (1997) have reported that adoption of VRT for fertilizers resulted in 57% higher profit compared to uniform application of similar quantity of nutrients. Bongiovanni and Lowenberg-DeBoer (2001) found that the computer model adopted to supply variable rates and statistical procedures adopted to arrive at appropriate rates of nutrients may also affect the economic advantages accrued. For example, use of spatial autocorrelation regression method was more accurate than ordinary least square for estimating crop response to nutrient supply. In other words, crop growth models used in the computer and economic analysis procedures are also crucial factors that may directly affect nutrients impinged, especially N and yield harvested.

Computer aided simulations depicting effects of fertilizer inputs have been very useful to maize farmers in the Midwest region of USA (Plate 1 and 2). They have been able to assess the effect of fertilizer dosages using VRT or even uniform rates on fields that are low or high in inherent soil fertility. Murat et al. (1999) report that positive effects of VRT were noticed at both low and high soil fertility regimes. In the low fertility field, profit from VRT was 7.8–27 US$ ha^{-1} and in high fertility regions, it was 7.5–58 US$ ha^{-1}. The maize crop yield improved if fertilizer input was enhanced.

Lowenberg-DeBoer and Aghib (1997) studied the economic advantages of applying precision farming with reference to P and K fertilizer inputs into cereal farms situated in Indiana, Ohio, and Michigan. The average net return was 357 US$ ha^{-1} for whole field management, 340 US$ ha^{-1} for grid soil sampling management and 370 US$ ha^{-1} for management zones based on soil type. The net profits may not be high, but adoption of precision farming clearly reduced the risk of poor grain yield and profits. The variability in net profits per farm was reduced considerable, if management zones were marked based on soil types and fertility zones. Rapid spread of precision farming depends immensely on the net profits that farmers accrue. Currently, cost of remote sensing and providing images that could be used in decision making within farms seems to offset the advantages. Satellite companies in USA and Canada charge

1.2 US$ ha^{-1} to create a readable NDVI vegetation map. They are generally derived from satellites such as Indian IRS-D, French SPOT or US Landsat. Commercial images of farming zones that could be utilized in precision farming costs 1.2–2.0 US$ ha^{-1} in Nebraska. In some locations, farmers have utilized low cost aerial photographs derived from parachutes or aircraft platforms. Perhaps, there is immediate need to subsidize costs on satellite imagery, if precision farming has to be more successful in crop belts that support production of fields crops of relatively low net value compared with cash crops.

PLATE 1 Top: Variable rate seeding in North Iowa-Corn belt.

Bottom: Seeding Monitors located in the Cabin of the planter helps farmers to regulate Seeding rates. Variable rate planting is quite common in the Corn Belt of North American plains.

Source: Mr. David Nelson, Nelson Farms Inc, Fort Dodge, Iowa, USA

PLATE 2 Top: A Combine Harvester fitted with GPS locator and "on-the-go" Yield Mapping Facility. Bottom: on-the-go loading of Maize grains from a Combine Harvester.
Source: Mr. David Nelson, Nelson Farms Inc, Fort Dodge, Iowa, USA

Equally so, farmers in dry lands who due to environmental vagaries need subsidies or free access to satellite imageries *via* computers and data recorders.

Long et al. (1996) have reported that in the wheat fields of Montana precision farming (VRT) is beneficial both in terms of nutrient supply to the cropping zone and economics. It saves on nutrient supply to the wheat fields. In addition, it improves profits. However, both nutrient dynamics, especially N and P, and productivity could differ. They are dependent on the sampling procedures and depth of soil profile considered while accruing soil data. Following is an example from wheat fields of Montana:

Depth of Soil Sampling	Cost of Fertilizer Technology US$ ha^{-1}	Profits US$ ha^{-1}
60 cm		
Variable Rate Technology	43	433
Uniform-Rates (Conventional)	17	372
120 cm		
Variable Rate Technology	25	344
Uniform Rates (Conventional)	10	327

Source: Long et al. 1996

It is said that soil test data that considers entire rooting depth including both surface and subsurface layers provides more accurate picture regarding soil N and P availability. It seems carry over-N or residual-N in the entire soil profile gets depicted better, if soil sampling is denser and deeper.

Let us consider a hypothetical case study that depicts how soil sampling procedures, VRT and crop response can affect economic feasibility of precision farming. These examples pertain to adoption of precision farming techniques in California and other regions in Southwest United States of America (University of California, 2010). In the first scenario, if soil sampling is done only at one place, it could fall in low or high fertility zone within the field or farm. If the soil sampling is from low yield potential zone, then fertilizer N or other nutrients stipulated will be marginally low. When low N is applied across the entire field, farmers tend to lose grain yield in the high yield potential zone. In the second scenario, if soil sample is from high grain yield potential zone, fertilizer-N stipulated in order to achieve high yield is proportionately high. This leads to loss of fertilizer-N or accumulation of unused fertilizer-N in the low yield zones. In the third scenario, if soil sampling is done sparsely in both potentially high and low grain yield zone. Small amounts of both grain and fertilizer could be lost. In the fourth scenario, if sufficiently extensive soil sampling has been done in the entire field or farm covering both high and low grain potential zones, then grain yield and fertilizer-N lost could be almost nil. The essence of considering such scenarios is that, extensive grid sampling that covers the entire field that receives variable rates of fertilizer-N or other nutrients is important. In other words, detailed sampling and appropriate understanding of the grain yield potential is necessary, if we intend to maximize economic benefits from precision farming. Computer model that determines the variable rate of nutrient release into soil is of course important.

Let us consider yet another example that deals with soil sampling procedure and VRT. In case of sugarbeet grown in North Dakota and Minnesota, Cattanach et al. (1996) found that management strips sampled using grid method and at two or three depths for soil-NO_3 provided better crop response. Grid sizes ranged from 1.0–1.6 ha and soil depths assessed for NO_3-N were 0–15 cm, 16–60 cm, and 61–105 cm. Fertilizer recommendations were based on 30–40 core samples. Economic analysis suggested that, if additional costs incurred on grid sampling and VRT were subtracted then net returns from precision techniques amounted to 143.0 US\$ ha^{-1}. It seems depth of soil sampling plays a crucial role in arriving at accurate N inputs, which are generally lower than if surface sampling is done.

Lowenberg-DeBoer (1995; 1999) studied fertilizer supply trends and economic benefits that accrue due to site-specific techniques. In most of the trials, input managed was primarily the fertilizer-based nutrients N, P, and K. Economically, 3 out of the 11 trials analyzed showed profitability, but almost all indicated that fertilizer input was much less compared to conventional whole farm prescription. This clearly indicates that although farmers aim at extra profits from site-specific techniques, we must realize that such techniques immensely affect nutrient dynamics, especially major nutrients. Reduction in N inputs has farfetched advantages with regard to agro-environment and soil quality, if entire cropping patch or agroecosystem is considered.

Field scale investigations spread across 18 locations in Missouri and Nebraska have clearly shown that, variable rate N inputs based on corn canopy reflectance during in-season, helps farmers in reducing N inputs. Firstly, it removes soil N variability and offers a degree of uniformity with regard to available N pools for corn roots to feed on. Fertilizer-N recovery and utilization improves vastly, if N supply is made accurately based on plant-N status and at exact intervals. Kitchen et al. (2009) have used crop-N sensors and variable rate N applicators and compared it with whole field N inputs for both economic and environmental impacts. Fertilizer-N inputs ranged from 0–235 kg N ha^{-1}. Generally, soil type, fertilizer cost, and corn price all affected economic advantages accruable from variable-N inputs. Yet, a modest profit of US\$ 25–50 ha^{-1} was achieved using variable N inputs. Fertilizer-N savings ranged from 10–50 kg N ha^{-1}. These reports clearly suggest that variable rate N inputs affect both N dynamics and productivity during maize cultivation. A report by Roberts et al. (2010) states that variable rate N inputs based on active light reflectance sensors are currently used extensively in the "Corn Belt of USA". It has been found beneficial in deciding in-season split application of fertilizer-N. Sensor-based supply of split N has immediate impact on N dynamics. It improves N recovery during active growth phase that is V7–11 stage. The net savings in fertilizer-N inputs and economic benefits from in-season N management using sensors and variable applicators have been generally significant compared to N rich controls or whole field applications.

In the maize belt of USA, more precisely in Illinois, farmers understand that increased profit from precision farming is mainly due to "input re-allocation" (Finck, 1998). Precision farming allows matching nutrient inputs with demand and with due consideration to weather pattern, soil fertility status and yield potential of crop genotype.

The net returns derived from conventional farming and precision farming is as follows:

	Net Returns (US$ ha^{-1})		
Crop	**Conventional**	**Manual**	**GPS**
Corn-Soybean 3 years	762	771	797
Corn 3 years	520	750	750
Soybean 3 years	133	138	135

Source: Finck, 1998

Note: Cost of adopting variable rate techniques on maize was 25 US$ ha^{-1} under manual and 49 US$ ha^{-1} under GPS guided systems. Net advantages due to precision farming over conventional systems ranged from 35–50 US$ ha^{-1}. Manual = Manual distribution of fertilizers. The GPS = GPS guided variable rate supply of fertilizers.

Kitchen et al. (2010) opine that corn farmers in USA prefer crop-N sensor techniques and variable rate N inputs because of increasing costs of fertilizer-N. They are all aiming at placing exact quantities of N accurately within each spot so as to economize and improve profitability. The prescribed N input is 235 kg N ha^{-1} but use of sensor-N readings and variable rate applicators suggest that N supply actually fluctuates by 100 kg N ha^{-1} in different locations within a field. It results in reduction of fertilizer-N requirement. Depending on corn genotype, region and soil type, a modest 25–50 US$ per ha^{-1} could be saved by using crop-N sensors and variable rate applicators. Clearly, precision techniques affect interlinked aspects like nutrient dynamics, crop productivity as well as economic returns.

In the Plains of North Texas, cotton cultivation using State Agency Recommendations for fertilizer-N supply considers the soils; yield goals and previous data on crop response. Fertilizer-N envisaged is said to be optimized for the entire region. It overlooks within farm or field variations. On the contrary, precision farming approach takes detailed account of soil fertility variations in each field and prescribes variable amounts of N. We should note that crop response to fertilizer-N is also dependent on interaction between soil nutrients and soil moisture status (i.e. irrigation). Quite often economic gains are dependent on fertilizer-N x irrigation interactions. Yu et al. (2001) made series of field scale evaluations around Lamesa in North Texas and found that site-specific techniques improve fertilizer-N efficiency. Therefore, fertilizer-N requirement for a given yield goal reduces. It also leaves a certain amount of residual nutrient that could be utilized later. The net economic gains from precision farming in comparison to conventional optimum fertilizer-N supply methods is summarized a follows:

Irrigation level	Precision farming	Whole-Field Farming	Change
50% ET			
Fertilizer-N applied (kg ha^{-1} y^{-1})	54	51	3
Lint Yield (kg ha^{-1} y^{-1})	780	770	10
Net Revenue (US$ ha^{-1} y^{-1})	890	789	11
75% ET			
Fertilizer N applied (kg ha^{-1} y^{-1})	96	93	3
Lint Yield (kg ha^{-1} y^{-1})	994	987	7
Net Revenue (US$ ha^{-1} y^{-1})	995	987	10

Source: Yu et al. 2001

Note: Net revenue refers to profits derived fertilizer-N and irrigation cost.

Yu et al. (2001) state that, there is clear advantage in terms of soil N dynamics and net revenue, due to change over from conventional State Agency Recommendations to precision farming.

Survey conducted in the Mississippi cotton belt has indicated that most farmers use map-based precision techniques. A sizeable portion, about 61% use GPS guided rapid response variable applicators. Surprisingly not many use hand-held sensor-based methods. The GPS-GIS based variable applicators were used for all three major nutrients N, P, and K. If not all, almost 83% of farmers adopting variable rate fertilizer supply techniques reaped higher lint yield ranging from 11–270 kg lint ha^{-1}. Several other surveys have shown that farmers gain at least 110 kg lint ha^{-1}more, if they followed precision farming techniques (Banerjee and Martin, 2007).

In the potato growing regions of Idaho, decision support systems and VRT have been in use for over a decade now. Reports indicate that fertilizer savings from VRT is significant. The savings ranged from 12.8 to 514.5 US$ ha^{-1} compared to uniform rate applications (Hoskinson and Hess, 1999).

Brazilian coffee crops thrive on Haplud Oxisols found on gentle slopes of Sao Paulo state. Fertilizer supply is often based on traditional practices that may underestimate crop P and K requirements. Recently, precision techniques were adopted to supply P and K fertilizers. It was based on soil tests and previously prepared soil P and K fertility maps. Field trials using large plots of 6.4 ha showed that variable rate P and K supply improved coffee seed yield by 38% over conventional uniform rate fertilizer supply. Fertilizer saving due to variable rate techniques was 23% for P and 13% for K compared to uniform rate inputs (Molin et al., 2010).

Forecast about economic benefits from precision farming were not entirely positive. For example, Fenton (1998) had opined that first phase of the precision techniques that involves soil fertility assessment, identification of variability and preparation of soil maps were most useful to the European farmers. It allowed them to act appropriately and bring down the soil fertility variations. The economic advantages of variable rate applications that depended factors like intensity of cropping, value of the grain/commodity and market price fluctuations, were still obscure. However, during recent years precision techniques are being used routinely in many of the cropping regions of Europe.

Field trials conducted in Southern England using site-specific techniques has shown that fertilizer-N application using variable rate techniques are profitable, compared with uniform supply of nutrients (Godwin et al., 2001; Welsh et al., 1999). The net profits due to precision farming ranged from 22–25 GB£ ha^{-1}. Higher N inputs generally lead to better grain yield and economic advantages. Anselin et al. (2004) have reported that comparison of net returns from N application indicates that, benefit from VRT was modest. Agronomists recommendations yielded 415 US$ ha^{-1}, but adoption of VRT systems yielded 427 US$ ha^{-1}. Malzer et al. (1999) found that grain/forage yield response was greater in high fertility zones compared with low fertility strips. Spatial difference in efficiency with which the crop has utilized fertilizer-N, supplied *via* VRT is crucial. Often, factors like water resources, drainage and soil compaction affected crop response to fertilizer-N applied using VRT. Net interaction of the VRT and soil factors seems important.

A study dealing with cereal production in Great Britain has shown that adoption of precision farming costs 5–18 GB£ pounds ha^{-1} depending on the exact system selected by farmers. At the prevailing prices of grains/forage it easily resulted in a net profit of 22 GB£ ha^{-1} (Godwin et al. 2003). The economic benefits depended on size of farm and crop response. They have suggested that for precision farming to be profitable, variable N rate effects should be perceived by at least 30% of the field in question.

Precision farming has been evaluated in many of the European countries that support large-scale wheat cultivation on the plains. They have examined its effect on grain yield, quality and economic advantages. Let us consider an example from Loess plains in Germany. Farmers in this region are usually conversant with soil fertility aspects. However, Mayer-Aurich et al. (2010) state that farmers possess a vague idea regarding potential economic advantages of adopting site-specific techniques. They adopted an *in situ* approach to assess economic advantages accrued from strip grown wheat that was produced with fertilizer-N applied at different levels. EC and soil chemical analysis formed the basis for variable applicators. They computed geo-referenced grain/forage yield from plots and found that economic advantages were marginal, although grain productivity and protein quality was better due to precision farming. The major reason stated is that farms were situated in low fertility regions and inputs too were low. It has been often stated that economic advantages from variable techniques are accentuated under high fertility conditions. Also, economic advantages derived from site-specific techniques were dependent on location, its general topography and agro-climate. The economic value of crop is also important.

Wagner (1999) has summarized results from trials in Germany and reported that, despite several interfering factors, average profitability from precision farming was 402 € ha^{-1} for winter wheat, 243 € ha^{-1} for corn, 204 € ha^{-1} for winter rape, 1665 € ha^{-1} for potatoes, and 190 € ha^{-1} for sugarbeet.

In a different study dealing with economics of fertilizer-N inputs and wheat grain productivity, Mayer-Aurich et al. (2007) found that fertilizer-N optimum with precision techniques was always marginally or significantly less than that observed under uniform-N management methods. In fact, economic benefits accrued due to precision methods were largely attributable to lower levels of N supplied. It ranged from 3–37 € ha^{-1} based on year and price levels of inputs and wheat grain. Following is an example from German plains that depicts profits due to adoption of precision techniques and detailed subplot harvests:

	Nitrogen Supply to Fields at Economic Optimum (kg ha^{-1})		
Year	**Uniform**	**Site-Specific**	**Site-Specific N + Sub Plot harvest**
2004	258	250	220
Profits (€ ha^{-1})	821	831	831
2006	258	251	223
Profits (€ ha^{-1})	1105	1118	1188

Source: Mayer-Aurich et al. (2007)

Note: Sub plot harvest increases accuracy while yield mapping.

As stated earlier, in addition to economic profits, adoption of precision farming influences N dynamics in the subplot/field or even cropping area if extrapolated. Fertilizer-N inputs are minimized and undue accumulation of residual-N in soil profile is avoided. It lessens chances of ground water contamination. This is a clear example wherein economic decisions of farmers do influence nutrient dynamics.

In France and neighboring wheat producing regions, satellite-based digital imagery has been regularly used to decide cropping pattern, especially wheat-based cropping systems. Fertilizer recommendations based on soil fertility maps and computer-based decision support systems are supplied to farmers/cooperatives at a cost. Adoption of precision farming based on remote sensed images has been profitable. For example, use of SPOT images and variable rate applicators reduced fertilizer-N requirement by 30 kg ha^{-1}, improved wheat yield by 200 kg ha^{-1} and increased gross profits by 53 € ha^{-1} (Astrium, 2002). It seems, farmers, it seems are most impressed by reduction in N supply to wheat fields. In addition to economic advantages, remote sensing with SPOT imagery helped in curtailing undue accumulation of nutrients in the soil profile. Based on simulations and field trials in Haute-Normandie, in France, Bourgain and LLorens (2009) have deduced that modern techniques like precision farming could be useful to improve N, P, and K dynamics in fields. It provides greater importance to environmental concerns and profits compared to previous procedures. Fertilizer (N, P and K) distribution could be regulated accurately to remove variations. Yet, we should realize that benefits of precision farming regarding nutrient dynamics, fertilizer efficiency and economic profits are all dependent on the scale of enterprise. They evaluated farms of the size 95 ha, 145 ha, and 240 ha and found that profitability was greater on larger sized farms and those with greater degree of soil fertility variability.

Economic analysis of variable rate techniques applied on maize grown at Bothaville, in South Africa has shown that gross margin was higher in management zones treated with VRT compared to whole fields treated with similar quantities of N and P. We should note that crop response to fertilizer-P supply might take effect for prolonged period. It is due to residual effect in soil. This phenomenon could make VRT more beneficial. Generally, economic benefits from precision farming improved during 2nd and 3rd year of adoption of VRT, since management zones take effect more consistently (Maine et al., 2005).

Cultivation of pearl millet then wheat in sequence is common to Western part of Gangetic plains, especially in the states of Rajasthan, Punjab and Haryana. Long-term field trials have consistently shown that nutrient depletion from the Inceptisols of this region is erratic and rampant. Therefore, fertilizer-based nutrient supply is kept relatively high. Precision farming (SSNM) approaches have been evaluated on both the crops. Results indicate that net returns from pearl millet under SSNM is INR 8,797 ha^{-1} compared to INR 4, 176 ha^{-1} under State Agency Recommendations. Net return per 1.0 INR invested was INR 2.10 for SSNM and INR 2.04 for SAR. The impact of SSNM was promising on wheat. Net return from wheat was INR 25,389 ha^{-1} for fields under SSNM and INR 16,652 ha^{-1} for those under SAR. The net return per INR 1.0 invested was INR 14.2 with SSNM and INR 6.7 with SAR.

INR = Indian Rupees

(Dwivedi et al., 2011). Introducing SSNM may be worthwhile. It may then bestow us with economic advantage plus delay the onset of soil deterioration. Fertilizer-N saved per season with wheat was 42 kg N ha^{-1} and 9.5 kg N ha^{-1} with pearl millet. Clearly, decisions made to economize, or improve profits have immediate effects on nutrient supply and other aspects like distribution, accumulation and ground water contamination.

Precision farming techniques have been evaluated on various vegetables, tubers and cash crops like sugarcane in the Southern Indian locations. In most cases, precision farming costs more than conventional systems. The extra cost involved depends on crop species, agronomic procedures, and yield goals. Reports by State Agricultural Agencies indicate that, although cost of cultivation increases, yield gain due to precision farming over conventional systems ranges from 39 to 200% for vegetables, 50% for tapioca, 33% for cotton, 37% for turmeric, 47% for bananas, and 67% for sugarcane. The agronomic procedures that actually cause higher costs need to be identified in each case. It may then guide us in identifying specific procedures that could be made economically more efficient. Following is an example of cost of cultivation and benefits from annual horticultural crops grown under conventional and precision farming systems in South India:

Agronomic Procedures/Inputs	**Conventional Systems**	**Costs** (INR ha^{-1}) **Precision farming**
Drip and Fertigation	15,000	
Field Preperation	5,600	7,500
Nursery and planting	6,000	12,000
Weeding	10,000	6,000
Plant Protection	10,000	8,500
Fertilizers	8,000	20,000
Wages	10,000	7,000
Staking	15000	
Total	**49,600**	**76,000**
Yield/expected yield	30 t ha^{-1}	135 t ha^{-1}

Source: India Development Gateway, 2010

Note: Costs shown pertain to 2004 and 2005. Approximately 47 INR = 1.0 US$

As stated earlier, precision farming involves a certain amount of higher fertilizer inputs and costs on agronomic procedures. Yet, it is beneficial because vegetable/fruit yield reaped is proportionately higher. A study by Maheswari et al. (2008) around Dharmapuri of Tamil Nadu in South India, suggests that farmers basically intend to modify nutrient dynamics through precision farming. They aim at better nutrient use efficiency and less detriment to agro-environment, especially to soil and water resources. At the same time, farmers adopting precision farming generally aim at higher vegetable yield. For example, in case of tomato, fruit yield improved by

80% compared to conventional systems. Similarly, brinjal fruit yield improved by 34% due to precision farming compared with conventional farming. There were also several constraints to adoption of precision farming. Such factors retarded spread of precision farming methods, despite clear advantages to agro-environment and farmer's economy. Among them, subsidy for cost of water-soluble fertilizers and irrigation were major constraints. Inadequate technical skills and low input technology adopted by dry land farms were other major constraints. The size of farm holdings is generally small and this may hinder rapid spread of precision technology in dry regions. The impact of economic factors on nutrient dynamics within the ecosystem needs due consideration.

Chinese cropping expanses include areas where both low and high input agriculture is practiced. Soils are highly variable with regard to soil fertility and precipitation pattern too is erratic in many regions. This allows ample scope for adoption of precision farming techniques that aim at overcoming within field variability in crop productivity. The advantages that accrue in terms of nutrient dynamics and grain/fruit yield may vary depending on input, technology adopted and yield goals envisaged. Let us consider an example. Maize is an important cereal that occupies both low and highly intensive cropping zones. Field investigations in Shanxi province, indicates that soils are variable for major nutrients N, P, and K as well as SOM distribution in the profile. Adoption of SSNM methods improved nutrient availability and imparted greater uniformity with regard to soil fertility and crop growth. Following is a summarized result derived from several fields in Shanxi province of China:

	Yield (kg ha^{-1})	**Net Income (US$ ha^{-1})**
Farmer's Practice	8355	1116
Site-Specific Nutrient Management	9380	1216

Source: Wang et al. (2006)

Economic benefits from SSNM have been investigated at field, farm and even village level. Reports by scientists at Soil Fertility Institute, CAAS, Beijing indicate that in a particular village (Ershilpu), wheat yield increased by 8–10 q ha^{-1} due to SSNM when compared with farmer's practice. Net increase in income averaged for 10 fields and 4 years between 2000–2004 was 5314 Yuan ha^{-1} (643 US$ ha^{-1}). It is said that better fertilizer recovery under SSNM was the major cause of higher grain yield and profits (Jiyun and Cheng, (2011). Several other trials conducted by International Plant Nutrition Institute in collaboration with Chinese Agricultural Agencies have shown that SSNM improved grain yields improved compared to Farmer's practices.

Crop	Yield Increase due to SSNM Over Farmer's Practice	
	kg ha^{-1}	%
Wheat	692	12.5
Maize	1086	16.8

Source: Jiyun and Cheng, 2011

Note: Yield increase depicted is average derived from 24 field trials for each crop between years 1998 and 2000; values in parenthesis denotes actual range of values.

Jiyun and Cheng (2011) suggest that enhanced fertilizer-N use efficiency was indeed a key factor for better performance under SSNM compared to farmer's practices. The fertilizer-N efficiency under SSNM increased by 7–8% over 30.3% that was already possible under farmer's practice. According to Guilian et al. (2003) fertilizer use efficiency in China is a trifle low at 30% compared to many developed western nations. Fertigation too has not improved fertilizer efficiency perceptibly. They believe that precision farming could improve fertilizer efficiency.

Wetland rice production is a major preoccupation in large tracts of South Asia. Rice production is often intensive and it involves wide range of electronic gadgets to regulate supply of chemical and organic manures. The SSNM was evaluated in over 170 locations within Asian rice belt. According to Dobermann et al. (2004), requirement of fertilizer-N decreased by 4% due to SSNM. Grain yield increases averaged 11% (0.1–0.6 t ha^{-1}) over conventional procedures. In the major rice growing regions of China, Southern India and Philippines, average increase in profits due to SSNM compared with conventional systems ranged from US$ 57 to US$ 82 per ha^{-1}crop season.

Tobacco culture has been steadily increasing in area in China. In addition, farmers have been revising the yield goals to higher level, in the hope that fertilizer efficiency and land productivity will improve. As such, need for fertilizer-based nutrients to support tobacco cropping zones has increased. However, regarding cash crops such as tobacco, it is believed that in China, adoption of traditional farming procedures has consistently underestimated the crop's need for nutrients. Therefore, precision farming approaches are being popularized in order to correct the mismatch of fertilizer supply and crop's demand for nutrients.

While highlighting the role of precision farming methods in terms of economic advantages possible to Australian farmers, Bramely (2006) states that it is based on certain facts and assumptions.

They are:

(a) Agricultural land, either individual field or a vast expanse is definitely variable with regard to factors like inherent soil fertility. In other words, mineral nutrient distribution, availability of nutrients supplied *via* fertilizers and/or organic manures, moisture, pH and aeration are variable;

(b) Rural landscapes comprise of flat, undulated or hilly tracts that are utilized for different purposes. They may possess drainage lines ranging from small creeks to large rivers into which water and nutrients may drain;

(c) Agricultural landscapes possess a range of soil types whose physico-chemical properties and productivity levels vary;

(d) Agricultural crop productivity may often, if not always varies with variation in soil fertility zones; and

(e) Australian farmers have known the facts since long. However, they did not know about precise techniques that allow them to measure and map variations in soil fertility and crop productivity. They were generally accustomed or forced to manage their fields using uniformly high fertility.

Further, according to Bramely (2006), key aspects that relate to usage of precision farming are:

(a) The extent of variation in soil factors;

(b) How and to what extent such variations relate to crop yield in quantity and quality;

(c) How accurately can we manage soil fertility variations by adopting precision farming methods; and

(d) what is the extent of economic and environmental advantage that accrues, if and when precision farming methods are applied to a particular field or agricultural area.

According to GRDC (2010), precision farming has beneficial effects on soil fertility, plus it improves economic advantages to farmers in Australia. Following is an example:

Location	**Annual benefit due to adoption of Precision farming (Au \$ ha^{-1})**		
	Fertilizer Management	**Grain/forage Production**	**Other Benefits**
Western Australia	12	9	6
South Australia	8	7	13

Source: GRDC, 2010

Note: Other benefits refer to improved soil characteristics, reduced irrigation, reduced weeds and so on. Fertilizer management refers to economic benefits accruing due to reduced fertilizer inputs for plots under precision farming.

We should note that farmers are prone to decisions about use of precision farming based on economic advantages. For example, reports by GRDC (2010) state that, wheat farmers in South Australia save at least 25 Au$ ha^{-1} y^{-1} due to adoption of precision framing. Further, fertilizer-based nutrient supplies are immensely affected by economic considerations. Hence, in this case, economic considerations seem to outweigh and affect nutrient dynamics in a farming. We need to view this aspect carefully.

KEYWORDS

- **Agricultural land**
- **Precision farming**
- **Site-Specific methods**
- **Sugarbeet**
- **Sugarcane**
- **Variable rate technique**

REFERENCES

Aishah, A. W., Zauyah, S., Anaur, A. R., and Fauziah, C. I. Spatial variability of selected Chemical characteristics of Paddy soils in Sawah Sempadan, Selangor, Malaysia. *Malaysian Journal of Soil Science*, **14**, 27–39 (2010).

Anselin, l., Bongiavanni, R., and Lowenberg-DeBoer, J. A spatial econometric approach of Site-Specific Nitrogen Management in Corn Production. *American Journal of Agricultural Economics*, **86**, 675–687 (2004).

Arnall, D., May, J., Butchee, K., and Taylor, R. *Evaluation of Sensor based Nitrogen application in Producers fields*. International Annual Meetings of American Society of Agronomy, Long Beach, California, USA, 107(5), pp. 1–2 (2010), Retrieved from http://a-c-s.confex.com/ crops2010am /webprogram/Paper58587.html (January 4th, 2011).

Astrium. *Farmstar: Crop management with SPOT*, pp. 1–3 (2002), Retrieved from http://www.spotasia.com.sg/ web/sg/2625-precision-farming.php (April 25th, 2011).

Attanandana, T., Suwannarat, C., Vaaraslip, T., Kongton, S., Meesawat, R., Bunampol, Soitosong, K., Tipanuka, C., and Yost, R. S. NPK Fertilizer Management for Maize: Decision Aids and Test kits. *Thai Journal of Soil and Fertilizer*, **22**, 174–186 (2000).

Australian Center for Precision Agriculture (ACPA). *Five main processes for a Site Specific Management System*. University of Sydney, Sydney, Australia, pp. 91–101 (2005), Retrieved from www.usyd.edu.au/su/agric/acpa/pag.htm (December 15th, 2010).

Banerjee, S. and Martin, S. W. *Summary of Precision-farming practices and perceptions of Mississippi Cotton Producers*. Mississippi Agricultural and Forestry Experiment Station Bulletin No 1157, pp. 1–43 (2007).

Bernardi, A. C. C., Gimenez, l. M., Silva, C. A., and Machado, P. L. O. A. *Variable rate application of Potassium fertilizer for Soybean crop growth in a No-till system*, pp. 1–18 (2003), Retrieved from http://www.icpaonline.org/finalpdf/abstract_138.pdf (June 13th, 2003).

Berntsen, J., Thomsen, K., Schelde, K., Hansen, G. M., Knudsen, L., Broge, N., Hougaard, H., and Horfarter, R. Algorithms for Sensor-based re-distribution of Nitrogen fertilizer in winter Wheat. *Precision Agriculture*, **7**, 65–83 (2006).

Blackmore, S. *The role of Yield maps in Precision Farming*. National Soil Resources Institute, Cranefield University, Silsoe, United Kingdom (Doctoral Dissertation), p. 161 (2003).

Blaise, J. *Integrated Nutrient Management for high quality Fiber and Yield*. Central Institute of Cotton Research, Nagpur, India, pp. 1–24 (2006), Retrieved from http://tmc.cicr.org.in/PDF/22.1.pdf (June 30th, 2006).

Bongiovanni, R. and Lowenberg-DeBoer, J. *Precision Agriculture in Argentina*. Third Simposio Internacional de Agricultura de Precision, pp. 1–14 (2005), Retrieved from http://www.cnpms. embrapa.br/siap2005/palestras/SIAP3-Palestra_Bongiovanni_e_LDB.pdf (March 25th, 2011).

Bongiovanni, R. and Lowenberg-DeBoer, J. Nitrogen management in Corn using Site Specific crop response estimates from Spatial regression model. Proceedings of the 5th International Conference on Precision Agriculture, Minneapolis, USA, pp. 1–8 (2001).

Bourgain, O. and LLorens, J. M. *Methodology to estimate economic levels of profitability of Precision Agriculture: Simulation for Crop systems in Haute-Normandie*, pp. 1–10 (2009), Retrieved from http://www.efita.net/apps/accesbase/bindocload.asp?d=6490&t=o& identobj =YjkgtriS&uid=57305290&idk=1 (June 21, 2011).

Bramely, R. *Precision Agriculture: Profiting from Variation*. Ecosystems Sciences Division, Council of Scientific and Industrial Research Organization, Glen Osmond, Australia, pp. 1–3 (2006), Retrieved from http://www.csiro.au/science/PrecisionAgriculture.htm (January 28th, 2010).

Burton, E., Roberts, R., and Sleigh, D. *Spatial distribution of Precision farming Technologies in Tennessee*. Department of Agricultural Economics. University of Tennessee, Knoxville. TN, USA, Research Report No 5, pp. 1–24 (2000).

Byju, G. and Suchitra, C. S. Nutrient Management Strategies in Tropical Tuber crops. *Indian Journal of Fertilizers*, **7**, 98–113 (2011).

Cattanach, A. Franzen, D., and Smith, L. Grid Soil testing and Variable rate fertilizer application effects on Sugar beet Yield and Quality. *Proceedings of the 3rd International Conference on Precision Agriculture*. Minneapolis, USA, pp. 1033–1038 (1996).

Chung, S., Sudduth, K., Jung, Y., Hong, Y., and Jung, K. Estimation of Korean Paddy field soil properties suing Optical reflectance. In *Asabe Annual International Meeting Technical papers*. American Society of Biological Engineers Annual International Meeting. Providence, Rhode Island. Paper No. 083682, pp. 1–3 (2008), Retrieved from http://asae.frymulti.com/abc.asp?JID=5&AID=25021&CID=prov2008&T=2 (December 15th, 2010).

Claret, M. M., Urrutia, R. P., Ortega, R. B., Stanely, B. S., and Valderrama, V. N. Quantifying Nitrate leaching in irrigated Wheat with different Nitrogen fertilization strategies in an Alfisol. *Chilean Journal of Agricultural Science*, **71**, 148–156 (2011).

Daberkow, S. G. and McBride, W. D. Socioeconomic profiles of early adopters of Precision Technologies. *Journal of Agribusiness*, **16**, 151–168 (1998).

Deshmukh, A. K. *Response of chilli to Site-specific nutrient management through targeted yield approach Northern transition zone of Karnataka*. University of Agricultural Sciences, Dharwar, India, pp. 1–126 (2008), Retrieved from http://etd.uasd.edu/ft/th9724.pdf. (June 30th, 2011).

Dobermann, A., Blackmore, S., Cook, S. E., and Adamchuk, V. I. Precision Farming: Challenges and Future Directions. In *New Directions for a Diverse Planet*. Proceedings of the Fourth International Crop Science Congress, Brisbane, Australia, p. 19, (2004), Retrieved from www.cropscience.org.au (January 20th, 2011).

Dwivedi, B. S., Sing, D., Swarup, A., Yadav, R. L., Tiwari, K. N., Meena, M. C., and Yadav, K. S. On-Farm Evaluation of SSNM in Pearl millet-based Cropping systems on Alluvial soils. *Indian Journal of Fertilizer*, **7**, 20–28 (2011).

Fenton, J. P. *Precision Farming: On-farm Experience*. Proceedings of International Fertilizer Society, pp. 1–2 (1998), Retrieved from http://www.fertilizer-society.org/proseedings/uk/Prc426.HTM (June 12th, 2011).

Fiez, T. E., Miller, B. C., and Pan, W. L. Assessment of spatially variable nitrogen fertilizer management in winter wheat. *Journal of Production Agriculture*, **7**, 86–93 (1994).

Finck, C. Precision farming can pay its way. *Farm Journal*, **122**, 10–13 (1998).

Godwin, R. J., Earl, R., Taylor, J. C., Wood, G. A., Bradley, R. I., Welsh, J. P., Richards, T., Blackmore, B. S., Carver, M. J., Knight, S., and Welti, B. *Precision Farming of Cereal crops*. A five-year Experiment to develop Management guidelines. Home grown Cereals Authority Project no 267, p. 28 (2001).

Godwin, R. J., Richards, T. E., Wood, G. A., Welsh, J. P., and Knight, S. M. An Economic analysis of the potential for Precision farming. *Biosystems Engineering*, **84**, 533–545 (2003).

Grain Research and Development Corporation (GRDC). *Precision Agriculture-Fact sheet-How to put Precision Agriculture into practice*. Kingston, Australia, pp. 1–6 (2010), Retrieved from www.grdc.com.au (January 1st, 2011).

Guilian, M., Hongatao, J., and Qin, Z. *Agriculture and Precision Agriculture*. The Information Research Institute, Shangai Academy of Agricultural Sciences, Shangai, China, pp. 1–8 (2003), Retrieved from http://www.afita.org/files/web-structure/20110126174028_862349/20110126174028_862349_97.pdf (June 9th, 2011).

Hammond, M. W. Cost analysis of Variable Fertility Management of Phosphorus and Potassium for Potato production in Central Washington. In *Site Specific Management for Agricultural Systems*. American Society of Agronomy, Madison, WI, USA, pp. 213–219 (1993).

Hongprayoon, C. *The movement from Conventional Agricultural practices to Integrated Nutrient Management in Thailand*. FAO Corporate Document Repository, pp. 1–5 (2010), Retrieved from http://www.fao.org/docrep/010/ag120e/AG120E17.htm (April 25th, 2011).

Hoskinson, R. L. and Hess, J. R. Using Decision Support systems for Agriculture (DSS4AG) for wheat fertilization. *Proceedings of the 4th International Conference on Precision Agriculture*. American Society of Agronomy, Madison, WI, USA, pp. 1797–1806 (1999).

India Development Gateway. *Precision Farming*. Tamil Nadu Agricultural University, Coimbatore, India, pp. 1–3 (2010), Retrieved from http://www.tnau.ac.in/horcbe/hitechfld.swf. (December 15th, 2010).

Jhoty, I and Autrey, J. C. *Precision Agriculture-perspectives for the Mauritian Sugar Industry*. Mauritius Sugar Industry Research Institute, Mauritius, Bulletin 12, pp. 1–7 (2000).

Jintong, L., Hong, C., Gaodi, X., and Ninomiya, S. *Generality for Precision Agriculture and Practice in China*, (2010), Retrieved from http://www.afita.org/files/web_structure/20110126174028_862349/20110126174028_862349_95.pdf (June 21, 2011).

Jiyun, J. and Cheng, J. *Site Specific nutrient Management in China: IPNI-China program*. International Plant Nutrition Institute, Norcross, Georgia, USA, pp. 1–7 (2011), Retrieved from http://www.ipni.net/ppiweb/china.nsf/$webindex/27D05B2887D6B7EE482573AE00293448?opendocument.

Kasetsart University, *Site-Specific Nutrient Management*, pp. 1–7 (2006), Retrieved from http:://www.ssnm.agr. ku.ac.th/main/Know/E_SSNM.htm. (December 15th, 2010).

Kent, S., Kitchen, N., Sudduth, K., Scharf, P., and Palm, H. Alternatives to using a reference strip for Reflectance-based Nitrogen application in Corn. *Proceedings of International Conference on Precision Agriculture*. Abstracts, pp. 1–3 (2007).

Kitchen, N., Shahanan, J., Roberts, D., Scharf, P., Ferguson, R., and Adamchuk, V. Economic and Environmental Benefits of Canopy Sensing for Variable rate-N Corn fertilization. In *Proceedings of the American Society of Agricultural and Biological Engineers Annual International Meetings*. Reno, Nevada, USA, pp. 1–3 (2009), Retrieved from http://asae.frymulti.com/abstract.asp?aid=27259&t=1 (December 15th, 2010).

Kitchen, N. R., Sudduth, K. A., Drummond, S. T., Scharf, P. C., Palm, H. l., Roberts, D. F., and Vories, E. D. Ground-based Canopy reflectance sensing for Variable rate Corn fertilization. *Agronomy Journal*, **102**, 71–82 (2010).

Lambert, D. and Lowenberg-DeBoar, J. *Precision Agriculture Profitability Review*. Site Specific Management Centre, School of Agriculture, Purdue University, Lafayette, IN, USA, p. 154 (2000).

Lilleboe, D. Will it Pay?. *The Sugar Beet Grower*, **25**, 18–20 (1996).

Long, D. S, Carlson, G. R., and Nielsoen, G. A. Cost of Variable Rate application of Nitrogen and Phopshorus for Wheat production in Northern Montana. *Proceedings of the 3rd Interntaional Conference on Precision Agriculutre*. Minneapolis, MN, USA, pp. 1019–1032 (1996).

Lowenberg-DeBoer, J. *Economics of Precision Farming: Payoffs in the future*. Procedings of Conference on Precision Decisions, University of Urbana-Champaign, Illinois, Illinois, USA, pp. 56–58 (1995), Retrieved from http://www.agriculture.purdue.edu/SSMC/Frans/ecoomic_issue.html (December 7th, 2010).

Lowenberg-DeBoer, J. Precision Agriculture in Argentina. *Modern Agriculture*, **2**, 13–15 (1999).

Lowenberg-DeBoer, J. *Precision Agriculture in Argentina*. Red Agricultura de Precision. Cardoba, Republic of Argentina, (2000), Retrieved from www.cnpms.embrapa.br/.../SIAP3_Palestra_Bongiovanni_e_LDP.pdf (May 28th, 2011).

Lowenberg-Deboer, J. *Precision Framing or Convenience Farming*, pp. 1–32 (2003a), Retrieved from http://www. regional.org.au/au/asa/2003/i/6/lowenberg.htm, (March 23rd, 2011).

Lowenberg-deBoer, J. *Precision Farming in Europe*, pp. 1–3 (2003b), Retrieved from http://www.agriculture. purdue.edu/ssmc/newsletters/June03_PrecisionAgEurope.htm (June 10th, 2011).

Lowenberg-Deboer, J. *Economic of Variable rate Planting for Corn*, pp. 1–3 (2006), Retrieved from http://agrarias.tripod.com/precision-agriculture.htm (March 25, 2011).

Lowenberg-DeBoer, J. and Aghib, A. *Average Returns and Risk Characteristics of Site-Specific P & K Management: Eastern Corn belt On-Farm-Trial Results*. Staff Paper 97-2. Department of Agricultural Economics, Purdue University, West Lafayette, IN, USA (1997).

Lowenberg-DeBoer, J. and Griffin, T. W. *Potential for Precision farming adoption in Brazil*. Site-Specific Management Center News Letter, Purdue University, Lafayette, Indiana, USA, pp. 1–3 (2006).

Lowenberg-DeBoer, J. and Swinton, S. M. Economics of Site-Specific Management in Agricultural crops. In *Site Specific Management for Agricultural Systems*, ASA/CSSA/SSSA/, Madison, WI, pp. 369–396 (1997).

Mahavishnan, K., Prasad, M., and Bhanu Recka, K. Integrated Nutrient Management in Cotton-Sunflower cropping system soils of North India. *Journal of Tropical Agriculture*, **43**, 29–32 (2005).

Maheswari, R., Ashok, K. R., and Prahadeeswaran, M. Precision Farming Technology, Adoption Decisions and Productivity of Vegetables in Resource-poor Environments. *Agricultural Economics Research Review*, **21**, 415–424 (2008).

Maine, N. and Nell, W. T. *Strategic approach to the implementation of Precision Agriculture principles in Cash crop farming*, pp. 217–225 (2005) http://www.farmingsuccess.com /id126.htm (February 21st, 2011).

Maine, N., Nell, W. T., Alemu, Z. G., and Barker, C. Economic analysis of Nitrogen and Phosphorus application under Variable and Whole field strategies in the Bothaville district of South Africa. *Research Report of Department of Geography*. University of Free State, Bloemfontein, South Africa, p. 10 (2005), Retrieved from http://ideas.repec.org/a/ags/agreko/7047.html.

Malzer, G. L., Mamo, M., Mulla, D., Bell, J., Graff, T., Strock, J., Porter, P., Robert, P., Eash, N., Braum, S., and Dikici, H. Economic Benefits and Risks Associated with Site-Specific Nutrient Management. In *Proceedings of the 5th Site-Specific Farming workshop*. Fargo, North Dakota, USA, pp. 234–239 (1999).

Matela, N. The status of Precision farming in Cash Crop Production in South Africa. *Masters Dissertation*. Department of Agricultural Economics, University of Free State, Bloemfontein, South Africa. p. 126 (2001).

Mayer-Aurich, A. Gandorfer, M., and Wagner, P. *Economic potential of Site Specific Management of Wheat production with respect to grain quality*, pp. 1–6 (2007), Retrieved from www.efita.net/apps/accesbase/bindocload.asp?d=6261&t=0 (May 28th, 2011).

Mayer-Aurich, A., Griffin, T. W., Herbst, R., Giebel, A., and Muhamed, N. Spatial econometric analysis of field-scale Site-Specific Nitrogen fertilizer experiment on wheat (*Triticum aestivum*) yield and quality. *Computers and Electronics in Agriculture*, **74**, 73–79 (2010).

Molin, J. P., Motomiya, A. V. A., Frasson, F. R., Faulin, G. C., and Tosta, W. Test procedure Variable-rate fertilizer on coffee. *Acta Sciencia Agronomy*, **32**, 1–13 (2010).

Murat, I, Khanna, M., and Winter-Nelson, A. *Investment in Site-Specific Crop Management under uncertainty*. Proceedings of the Annual Meeting of American Agricultural Economics Association. Nashville, Tennessee, USA, pp. 78–79 (1999).

Norasma, C. Y. N., Shariff, A. R. M., Amin, A. S., Khairunniza-Bejo, S., and Mahmud, A. R. *Web-based GIS Decision Support System for Paddy Precision farming*. Web Precision Farmer V. 2.0 Universiti Putra Malaysia, Selangor, Malaysia, pp. 1–3 (November 29th, 2010) (2010).

Norton, G. W. and Swinton, S. M. Precision Agriculture. Global prospects and Environmental implications. In *Tomorrow's Agriculture Incentives, Institutions, Infrastructure and Innovations*. G. H. Peters and P. Pingali (Eds.). Aldershot, United Kingdom, pp. 73–84 (2001).

Pedersen, S. M. A multi-perspective report on application of advanced systems-economic, usability and acceptability. *Future Farm*, pp. 1–2 (2011), Retrieved from http://www.futurefarm.eu/nose/241 (June 10th, 2011).

Roberts, D., Shanahan, J., Richard, F., Adamchuk, V., and Kitchen, N. Integration of an Active sensor algorithm with Soil-based Management zones for Nitrogen management in Corn. *Proceedings of American Society of Agronomy Annual meetings*. New Orleans, USA, pp. 1–2 (2010), Retrieved from http://a-c-s.confex.com/crops/2010am/webprogram/paper59152.html (December 15th, 2010).

Roberts, R. K., English, B. C., Larson, J. A., Cochran, R. L., Goodman, B., Larkin, S., Marra, M., Martin, S., Reeves, J., and Shurley, D. *Precision farming by Cotton producers in six Southern states: Results from 2001 Southern Precision farming survey*. Department of Agricultural Economics, The University of Tennessee, Knoxville, Tennessee, USA Research Series 03–02, pp. 1–83 (2002).

Schepers, J. S. *Precision Farming: One key to Quality Water*, pp. 1–3 (1996), Retrieved from www.fluidfertilizer.com/pastart/pdf/13P28-31.pdf (May 28th, 2011).

Sparovek, G. and Schnug, A. Soil tillage and Precision Agriculture: A theoretical case study of soil erosion in Brazilian Sugarcane production. *Soil and tillage*, **61**, 47–54 (2001).

Srinivasa Rao, C. Nutrient management strategies in Rainfed Agriculture: Constraints and opportunities. *Indian Journal of Fertilizers*, **7**, 12–25 (2011).

Sudduth, K. A., Kitchen, N. R., Scharf, P., Palm, C., and Shannon, H. *Field-scale N application using Crop Reflectance Sensors*. American Society of Agronomy Annual Meetings Abstracts New Orleans, USA, Paper No 153–157, p. 1 (2007).

Sudduth, K., Newell, K., and Scott, D. Comparison of three Canopy Reflectance Sensors for Variable-rate Nitrogen application in Corn. *Proceedings of the International Conference on the Precision Agriculture*. Abstract, pp. 1–2 (2010), Retrieved from www.ars.usda.gov/ pandp/people/people.htm?personid=1471.html (December 15th, 2010).

Swinton, S. M. and Lowenberg-DeBoer, J. *Profitability of Site Specific Farming*. Site-Specific Management Guidelines, SSMG-3. Potash and Phosphate Institute, Norcross, Georgia, USA, p. 35 (1997).

Swinton, S. M. and Lowenberg-DeBoer, J. *Global adoption of Precision Agriculture Technologies: Who, When and Why*, pp. 557–562 (2005), Retrieved from https://www.msu.edu/user/swintons/D-7-8_SwintonECPAO.pdf (March 25th, 2011).

Sylvester-Bradley, R., Lord.E., Sparkes, D. L., Scott, R. K., and Williams, S. An analysis of the potential of Precision Farming in northern Europe. *Soil Use and Management*, **15**, 1–8 (1999).

Tran, D. V. and Nguyen, N. V. The concept and implementation of Precision Farming and Rice Integrated Crop Management systems for sustainable production in the twenty first century. *A report on Integrated Systems*. Food and Agricultural Research Organization of the United Nations, Rome, Italy, pp. 91–101 (2008).

University of California. *Precision Agriculture: Economics of Precision Agriculture*, pp. 1–22 (2010), Retrieved from http://www.precisionag.org/html/ch14.html (February 21st, 2011).

Wagner, P. *The Future of Precision Farming*. The Development of a Precision Farming information system and Economic aspects, pp. 1–13 (1999), Retrieved from http://www.lb.landw.uni-halle.de/publikationen/pf/pf_efita99.htm (June 2, 2011).

Wang, H., Jin, J., and Wang, B. Improvement of Soil Nutrient Management *via* information technology. *Better Crops*, **90**, 30–32 (2006).

Welsh, J. P., Wood, G. A. Godwin, R. J., Taylor, J. C., Earl, R., Blackmore, B. S., Sphor, G., Thomas, G., and Carver, M. *Developing strategies for spatially variable nitrogen application*. Second European Conference on Precision Agriculture Odense, Denmark, pp. 2–24 (1999).

Xie, G., Chen, S., Qi, W., Lu, Y., Yang, X., and Liu, C. A multidisciplinary and Integrated study of Rice Precision Farming. *Chinese Geographical Science*, **13**, 9–14 (2007).

Yu, M., Segarra, E., Watson, S., Li, H., and Lascano, R. J. Precision Farming Practices in irrigated Cotton production in the Texas High plains. *Proceedings of the Beltwide Cotton Conference*, **1**, 201–208 (2001).

5 Precision Farming: Summary and Future Course

CONTENTS

Precision is an important concept that needs greater attention during adoption of almost every farming practice, beginning from land preparation, seeding, fertilizer supply to harvesting, and grading of the produce. Historical records on crop production trends from different continents indicate that imparting greater precision has often improved crop productivity. There has been a constant improvement in instruments and techniques that add precision during crop production. During recent years, there has been a spurt in soil testing, nutrient analysis and use of soil fertility maps, so that variable rates of fertilizers could be dispensed. Hand-held or in situ sensors and remote sensing techniques that aid in soil fertility and grain yield mapping have again improved precision. Rapid "on-the-go" sensors and computer guided decision support systems have revolutionized fertilizer and irrigation prescription.

It is generally believed that precision techniques are amenable and economically more viable, if the farms are large, soil fertility variation is significant, and economic value of the produce too is relatively higher. This might be true in a couple of continents where large scale farming is in vogue. For example, in North America, adoption of precision techniques helps in reducing soil fertility variations. It reduces fertilizer and chemical inputs to farms. Reports suggest that fertilizer-N supply could be reduced perceptibly. Regarding soil N and fertilizer-N management using precision farming approaches, Doerge (2011) opines that right now, in United States of America, number of farmers opting for N management using variable rate technology is still small. Field demonstrations that prove the advantages precision techniques over uniform fertilizer supply needs greater attention. Understanding temporal variations in soil N availability and its use during variable rate application of fertilizer-N seems pertinent. Further, it has been stated that fertilizer-N recommendations do not consider economically optimum rates. However, we really do not know whether economics and grain price fluctuations should be considered, while regulating a soil nutrient and crop physiological phenomenon. We have no idea about long term effects of trying to regulate natural phenomena based on market pricing and whimsical fluctuations in grain prices, if it occurs. Several countries use different levels of subsidies on fertilizer and grain pricing. Their effects may or may not be congenial, if fertilizer-N supply is regulated entirely by profit margins. Precision farming should bestow greater importance to natural consequences of fertilizers and chemicals on soils, agro-environment and crop productivity. Economic consequences could be dealt suitably through administrative and legislative policies.

Precision technique involves a few different aspects like soil analysis, mapping, decision support system, variable rate applicators and yield monitoring. Farmers are shrewd and they do not adopt all of these aspects of precision farming in all the agroecoregions. For example, in Argentina, soil fertility variations are small and fertilizer supply levels are marginal or medium. This situation does not offer great advantages, if elaborate soil sampling and variable techniques are adopted. However, a good knowledge of variations in soil fertility productivity is essential. Hence, Argentina farmers have preferred to buy and use vehicles fitted with yield monitors and yield mapping devises. Similarly, in India and other Southeast Asian regions, farms are small and economic value of produce too is not high enough to buy sophisticated GPS guided vehicles and variable rate fertilizer dispensers. In fact, large scale sampling, massive equipments and variable rate techniques become economically non viable. They need to be augmented through larger cooperatives. Hence, most farmers, currently, prefer use of cheaper, hand-held sensor devices that help in mapping soil fertility, crop N status and grain yield. In some countries, soil or crop maps are made affordable through remote sensing agencies. In regions that support only subsistence farming, adoptions of elaborate techniques are definitely not remunerative. However, naturally, subsistence farms too show up a great degree of soil fertility variations. Crop growth and grain/forage yield trends too vary conspicuously. In many parts of Sub-Sahara, the undulated sandy terrain causes soil fertility variations and uneven crop growth, which is generally termed as a natural phenomenon. Again, suitable agricultural extension programs, supply of accurate soil fertility/crop growth maps derived from remote satellites and matching fertilizers may help in improving crop productivity. This region obviously needs large subsidies and suitably modified precision farming approaches. In subsistence farming regions, major advantages from precision techniques are confined to reduction in fertilizer consumption, reduction in loss of precious soil nutrients and marginal improvement in grain yield.

Globally, we up-regulated cropping intensity by developing high-yielding genotypes, fertilizer technology and water resources. Several factors affected spread and intensity with which a crop or its genotype found its way into different corners of an agroecosystem. The supply of fertilizer-based nutrients into cropping zones often matched the crop demand and yield goals. Economic resources and profit margins too dictated the nutrient dynamics and net production of biomass in a region. Whatever is the reason, fertilizers impinged into ecosystems increased rather drastically and sometimes rampantly affecting the environment?

We should note that, in certain regions, incessant intensive cropping has lead to application of fertilizers and other chemicals, uncontrollably in quantities more than crop's demand. The undue accumulations, loss via seepage, erosion, percolation, and emissions became conspicuous. Soil fertility variations too got magnified. In many regions it almost deteriorated the soil and affected ecosystem. Precision farming indeed has an important role in intensive farming zones.

Precision farming thwarts soil deterioration by regulating supply of fertilizer and other chemicals. Perhaps precision farming is an important technique that improves agronomic efficiency of fertilizers. We can regulate the fertilizer-based nutrient supply, chemical sprays, and irrigation more accurately and at desired levels. This is a great advantage, since it allows us to delay or altogether reduce soil and environmental deterioration. We can regulate nutrient transformation rates, its loss and recycling patterns. To a certain extent, nutrient turnover in the field could be regulated through precision techniques.

The spread of precision farming techniques have been relatively slow in many parts of world including North America and Europe. According to Ess (2002) crux of the problem is in our ability to convince that precision techniques are easy to handle and economically feasible and even profitable. Perhaps, there is a need to highlight its advantages in terms of accuracy, sophistication, grain yield improvement, fertilizer use efficiency, and environmental advantages. Ess (2002), further adds that slow pace of spread of precision farming is attributable to interdisciplinary nature of precision farming. Researchers at Site-Specific Crop Management Center of University of Purdue University, Indiana in USA, opine that there are not many private agencies that pool wide array of specialists like engineers to control electronically guided traction and spray machinery, seed specialists, agronomists, irrigation engineers, entomologists to orchestrate a precision farming process. Currently, only governmental agencies and universities are applying their personnel to work out precision techniques. Private companies may have to invest at higher levels to build up a team. Despite, problems related to scale of investment, precision farming approach would be plausible. They would become routine in many agrarian zones, where large farming enterprises thrive, state farms are maintained and cooperatives that afford to pool resources and technology are situated. We should also realize that many of the techniques that are still in their rudimentary stages of sophistication and are costly, but would eventually become easy to afford and use.

On a time scale, precision farming could be a good chance and right opportunity for policy makers dealing with agricultural expanses and ecosystems to regulate nutrient dynamics and productivity levels. Precision farming may just follow the same course traversed by variety of other technologies that initially were difficult to handle, practice, master, and economically not feasible. However, sooner or later appropriate modifications were effected equipments and procedures became economically more feasible. We should be alert to the fact that, irrespective of economic gains, precision farming should be adopted just because it regulates soil fertility, various natural process and productivity in a given agro-environment. Precision techniques offer to protect soil fertility, regulate nutrient dynamics, and several other natural phenomena without marked effects on crop productivity. Hence, in future, farmers may prefer precision techniques more frequently.

Overall, precision techniques seem to have great future in all agrarian regions of the world. Precision techniques, as they get refined and become routine will lead us to "Push Button Agriculture". Agricultural operations get highly sophisticated, electronically controlled, and excessively more convenient to practice. Most, if not all agricultural operations could then be carried out at an appropriate pace and accurately. This aspect is an important incentive to every farmer in all regions of the world, irrespective of economic gains. Robotics, rapid soil analysis, and computer-based decision support systems need to be improved. This is crucial because it bestows farmers with greater versatility, so much that they can revise and re-revise farm operations in a crop season, with ease to suit environmental vagaries, yield goals, and economic forecasts.

In future, remotely placed computers could be regulating farm operations precisely. Literally, computers (supervisors) stationed in all together different country or continent could be deciding, regulating precisely all farm operations and reaping harvests.

KEYWORDS

- **Intensive farming zone**
- **Precision farming**
- **Precision technique**
- **Push button agriculture**
- **Sensor devices**

REFERENCES

Doerge, T. A. *Variable rate Nitrogen Management for Corn production success proves elusive*. International Plant Nutrition Institute, Norcross, Georgia, USA, pp. 1–2 (2011).

Ess, D. R. *Precision and Profits: Producers heed bottom line when considering high technology farming methods*. Resource Engineering and Technology, pp. 1–3 (2002), Retrieved from http://www.highbeam.com/doc/IGI-83374244.html.

Index

P

R

S

Milton Keynes UK
Ingram Content Group UK Ltd.
UKHW031150141024
449569UK00024B/904

9 781774 632703